Longman Science for AQA

GCSE Science Year 10

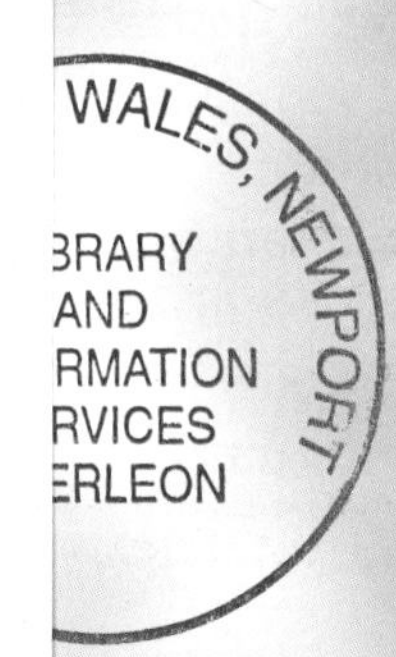

Miles Hudson
Penny Johnson
Sue Kearsey
Colin Lever
Penny Marshall

Edinburgh Gate
Harlow, Essex

How to use this book

The book is divided into 6 units. Each unit has a one-page introduction and is then divided into topics. At the end of each unit there is a set of questions that will help you practise for your exams, some practice coursework questions and a glossary of key words for the unit.

As well as the paper version of the book there is a CD-ROM called an ActiveBook. For more information on the ActiveBook please see the next two pages.

What to look for on the pages of this book:

Learning objectives
These tell you what you should know after you have studied the topic.

Glossary words
You will need to know the meaning of some key words. These are shown in **bold**. The glossary at the end of each unit gives you a list of all the key words and what they mean.

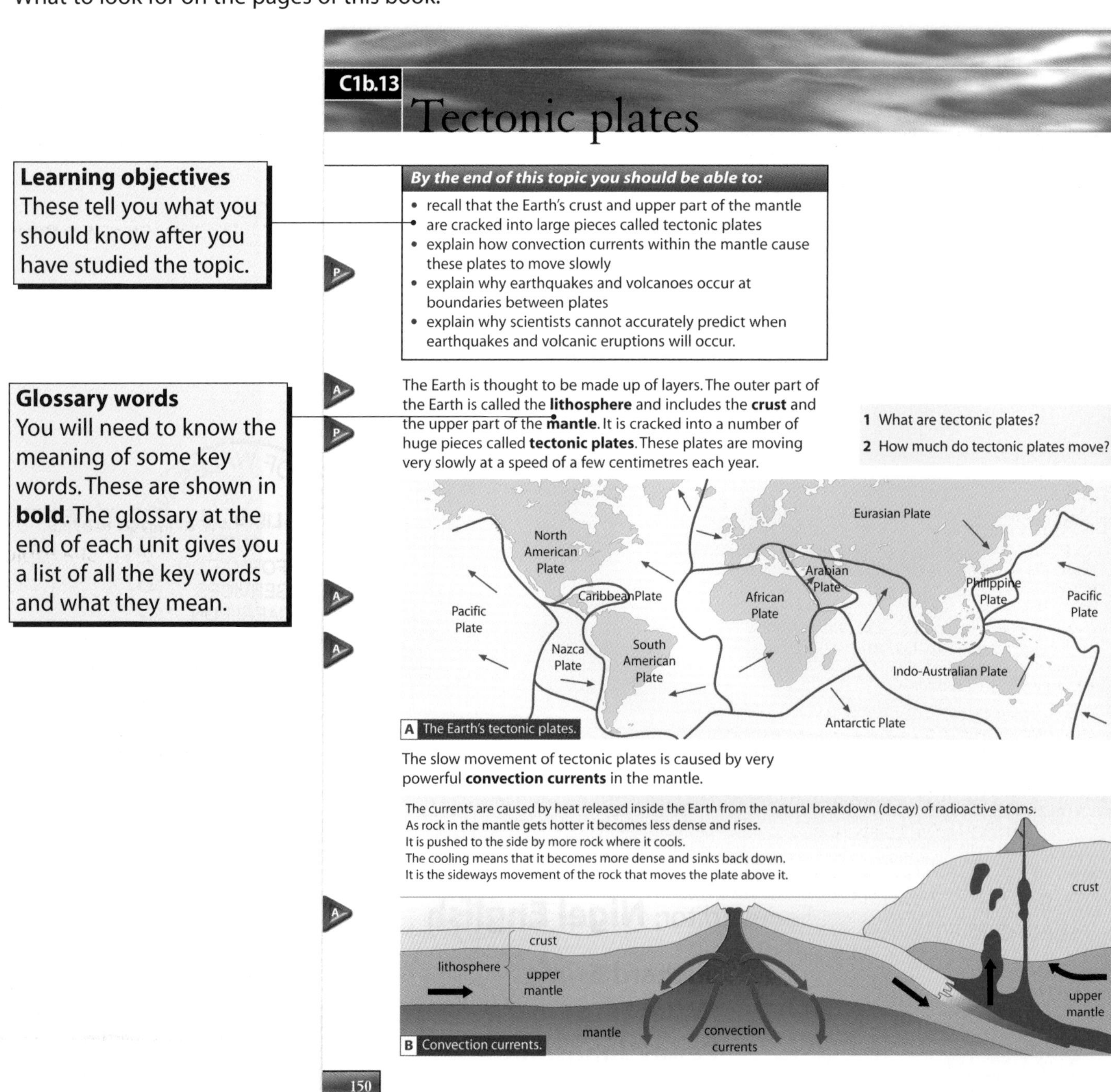

Volcanoes and **earthquakes** occur at the boundaries between tectonic plates. As plates move past, over, under or apart from each other, hot magma from the mantle can escape, resulting in a volcano. Friction between the moving plates may stop the plates moving smoothly. If this happens, the plates move in sudden jerks resulting in earthquakes.

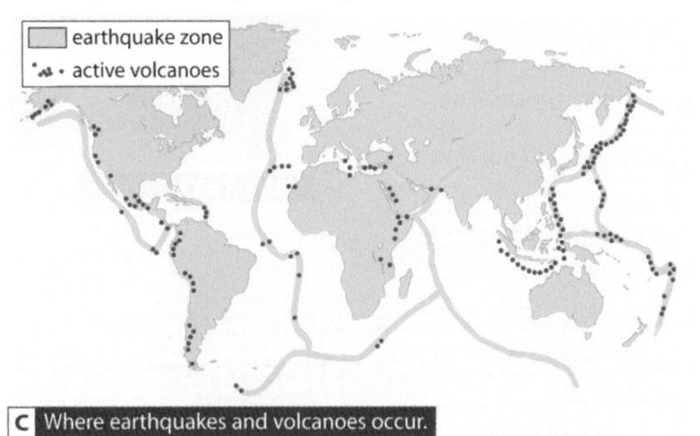

C Where earthquakes and volcanoes occur.

3 Explain why tectonic plates move.

4 What causes convection currents inside the Earth?

5 Why do volcanoes and earthquakes develop at plate boundaries?

Questions
There are lots of questions to help you think about the main points in each topic.

Predicting earthquakes and volcanic eruptions

Many people live in areas at risk from earthquakes. Scientists know where earthquakes are likely and so many buildings in danger zones are built with special foundations to help them to withstand earthquakes.

There are some warning signs before an earthquake:

- small earthquakes (pre-shocks)
- water levels in wells fall
- some animals act strangely.

Volcanic eruptions are much easier to predict than earthquakes. Warning signs before an eruption include:

- increasing temperature of the volcano due to magma moving underground
- rising ground level around the volcano due to the build up of magma
- more sulfur dioxide (SO_2) gas is given out.

When these warning signs appear, people can be moved to safety. However, scientists cannot reliably predict exactly *when* major earthquakes or volcanic eruptions are going to happen. It is difficult to predict when there will be enough pressure for plates to slide past each other or for magma to burst up.

6 Why is it important to be able to predict where and when earthquakes and volcanic eruptions will happen?

7 Which is easier to predict: an earthquake or a volcanic eruption? Explain your answer.

8 Why can scientists not reliably predict exactly when earthquakes or volcanic eruptions will occur?

9 Imagine that you are a journalist living in an earthquake zone near a volcano. Write a short newspaper article (no longer than 200 words) explaining:

a why earthquakes and volcanoes occur where you live

b what scientists can do to predict earthquakes and volcanic eruptions.

How science works
Some sections of a topic might focus on science in everyday life or the practical skills involved in science. These are highlighted where they appear in each topic.

Summary exercise
The answer to the last question in the topic summarizes the whole of the topic. The exercise will also be useful for revision.

151

How to use your ActiveBook

The ActiveBook is an electronic copy of the book, which you can use on a compatible computer. The CD-ROM will only play while the disc is in the computer. The ActiveBook has these features:

Glossary
Click this tab to see all of the key words and what they mean. You can read them or you can click 'play' and listen to someone else read them out for you to help with the pronunciation.

DigiList
Click on this tab and all the electronic files on the ActiveBook will be listed in menus.

ActiveBook tab
Click this tab to access the electronic copy of the book.

Key words
Click on any of the words in **bold** to see a box with the word and what it means. You can read it or you can click 'play' and listen to someone else read it out for you to help with the pronunciation.

Interactive view
Click this button to see all the bits on the page that link to electronic files. You have access to all of the features that are useful for you to use at home on your own. If you don't want to see these links you can return to **Book view**.

ActiveBook DigiList Glossary

C1b.14

The theory of continental drift

HSW *By the end of this topic you should be able to:*

- explain what the theory of continental drift is
- explain why it took many years for the theory of continental drift to be accepted.

As apples get older they start to dry up and shrink. This makes the skin too big and so it wrinkles up. For many years it was believed that the features on the Earth's surface were made in a similar way. As the young, hot Earth cooled, the crust shrank and wrinkled, forming mountains.

A Apples shrink and wrinkle.

1 How did people believe mountains were formed before the theory of plate tectonics?

In 1911 Alfred Wegener read that the fossils of identical creatures had been found in South America and Africa, and that these creatures could not swim. People at the time said that there must have been a piece of land between the continents which was now covered by the Atlantic Ocean.

Wegener came up with the idea of **continental drift**. He believed that the continents were moving around and were once joined together in a big land mass. In 1915, Wegener published a book about his theory.

The main evidence in Wegener's book was:

- the continents appear to fit together like a jigsaw
- the west coast of Africa and the east coast of south America have the same patterns of rock layers.
- these two coasts have the same types of plant and animal fossils; some of these animals are only found in those parts of the world and their fossils show they could not swim
- there could not have been a piece of land connecting South America and Africa, which means they must have been joined together.

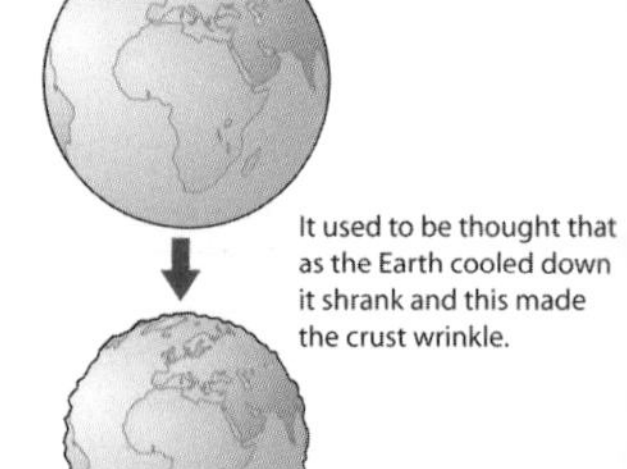

It used to be thought that as the Earth cooled down it shrank and this made the crust wrinkle.

B The shrinking Earth theory.

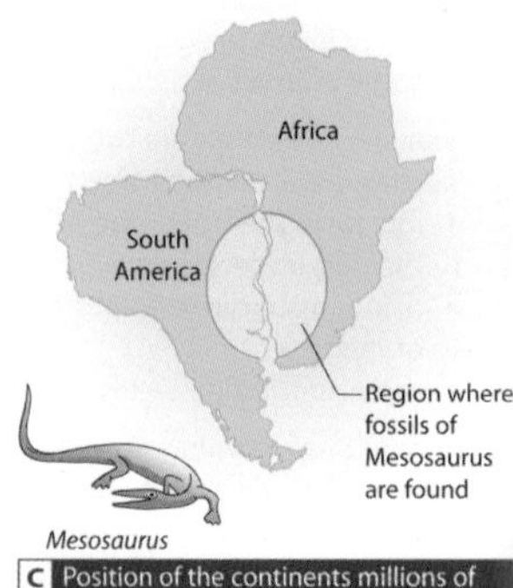

C Position of the continents millions of years ago and the region in which fossils of *Mesosaurus* have been found.

152

© Pearson Education 2006 Turn off

Go Interactive

Target sheets
Click on this tab to see a target sheet for each unit. Save the target sheet on your computer and you can fill it in on screen. At the end of the topic you can go back and see how much you have learnt by updating the sheet.

Help
Click on this tab at any time to search for help on how to use the ActiveBook.

sheets Help

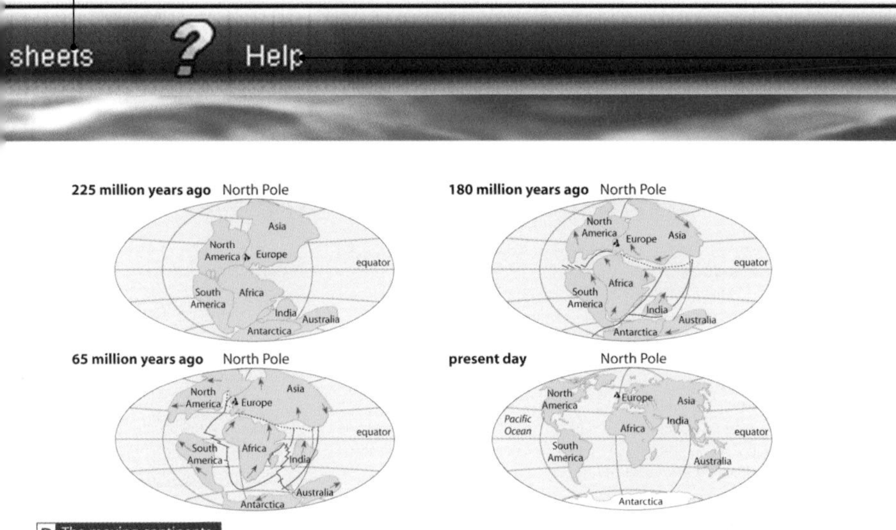

D The moving continents.

When Wegener died in 1930 his ideas had still not been accepted. This was because there was no explanation for how the continents moved and because Wegener was not a geologist.

His ideas were not accepted until the 1960s when the Atlantic Ocean floor was surveyed in detail and the mid-Atlantic ridge was found. This is a range of underwater mountains and volcanoes in the middle of the ocean.

Soon afterwards it was discovered that the rock in the ocean floor is younger than the rock in the continents. The rock closest to the ridge is youngest. It was also found that the magnetic alignment of rocks containing iron-rich minerals was symmetrical either side of the ridge. This new evidence fitted in with the theory of continental drift.

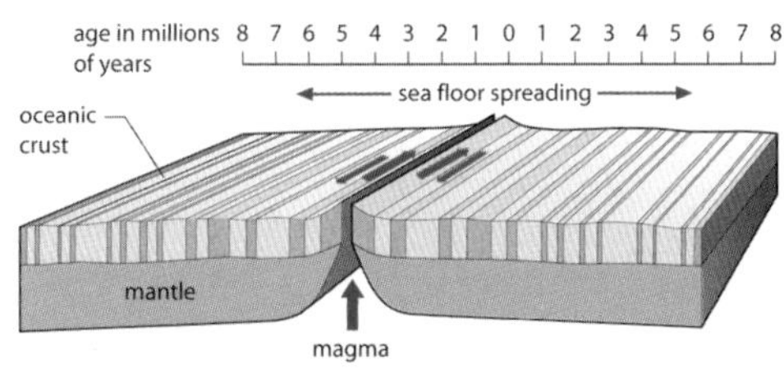

E Sea floor spreading. This is caused by the Earth's magnetic field reversing its direction every few hundred thousand years.

Scientists also discovered that this movement could be caused by very powerful convection currents in the mantle. This brought about the new theory of plate tectonics which is very similar to Wegener's theory of continental drift.

2 What is meant by continental drift?
3 How do rock patterns support Wegener's ideas?
4 How do fossils of *Mesosaurus* support Wegener's ideas?
5 What other evidence did Wegener use to support his ideas?

6 Give two reasons why Wegener's ideas were not accepted at the time.
7 What evidence was discovered that led to Wegener's ideas being accepted?

8 Imagine you were a scientist living in the 1960s. List all the evidence you have to support the theory of plate tectonics.

153

of 255

Zoom feature
Just click on a section of the page and it will magnify so that you can read it easily on screen. This also means that you can look closely at photos and diagrams.

Page number
You can turn one page at a time, or you can type in the number of the page you want and go straight to that page.

Contents

Human biology

A Iwan Thomas warming up.

Everything affects your body: from the food that you eat to what you do in your leisure time. Your body senses changes that take place and responds to try to keep you healthy.

We all need to eat a healthy and balanced diet, exercise regularly and avoid taking things that might cause us harm. Athletes take even greater care of their bodies. Taking care of what and when they eat and drink helps to speed up recovery time and protects them from illness and injury. Training not only makes them fit, it also helps to improve their reaction times.

By the end of this unit you should:

- be able to explain how your nervous system helps you to respond to your environment
- be able to explain how our body regulates body temperature, blood glucose and water
- understand the action of hormones and the role they play in fertility
- be able to evaluate the effect of food on health
- know how drug misuse affects the body
- know about the causes of disease and how your body protects you against infection
- be able to describe the development of medicines to help protect you from disease.

1 What makes Iwan Thomas faster than other athletes?

2 What has he done to make himself quicker?

3 a Why are some people not fit and healthy like an athlete?
b How does this affect their body?

The nervous system

By the end of this topic you should be able to:

- explain that receptors detect stimuli
- describe how your nervous system enables you to react to your surroundings and co-ordinates your behaviour
- explain the pathway from stimulus to response
- describe what a reflex action is.

The nervous system is made up of three main parts: the brain, the spinal cord and nerve fibres. It detects **stimuli** such as light, temperature, sound, pressure, pain, movement, touch and chemicals (for example, food) and co-ordinates the body's response. If you smell something burning your nose sends a message to your brain, and it co-ordinates your response.

1 Look at photograph A. Explain, in detail what happens the moment the starter fires the gun.

A The difference between winning and losing a race might be down to how quickly you start. It is all about **reaction time**. This is the time between recognising the signal and reacting to it.

B The nervous system.

Receptors are sensors on the body that detect stimuli. Information from these passes along nerve fibres to the brain. Nerve fibres are made from lots of nerve cells, called **neurones**. The information passes along them as electrical signals called **impulses**.

There are three types of neurone:

- **Sensory neurones** carry impulses from the receptor to the spinal cord.
- **Relay neurones** carry impulses through the spinal cord and up to the brain and from the brain back along the spinal cord.
- **Motor neurones** take impulses from the spinal cord to an **effector**. An effector can be a muscle that is made to contract (tighten) or a gland that releases (secretes) a chemical (for example, a hormone).

Neurones are not joined to each other. There is a small gap between them called a **synapse**. When an impulse reaches the end of a neurone, a chemical is released. This travels across the gap, and starts an impulse in the next neurone. This means the message can only travel in one direction.

Usually the brain checks all the impulses and decides how to react, but sometimes a quicker response is required. For example, if you accidentally touch something hot, you pull your hand away very quickly.

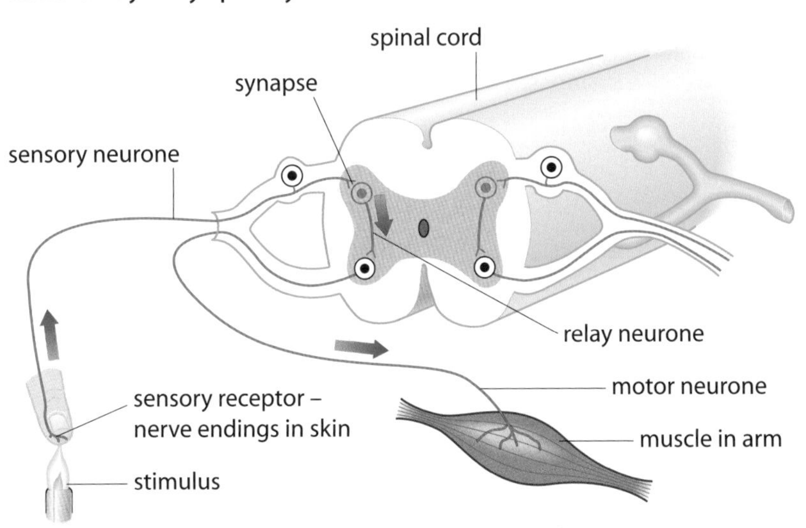

D What happens when you touch something hot?

Look at diagram D. The heat is detected by a sensory receptor in your finger. It sends an impulse along a sensory neurone to a relay neurone in your spinal cord. Instead of sending a message to your brain, it sends an impulse via a motor neurone to an effector, a muscle, which contracts, pulling your hand away. This is called a **spinal reflex**, or **reflex arc**. It is automatic and very quick. There are other types of reflex actions such as coughing and blinking. These are needed to protect us from being hurt.

2 **a** Identify six stimuli that you respond to.
b Where are the sensory neurones that respond to each of these stimuli found?

3 Some people suffer from a disease called motor neurone disease.
a Suggest which part of the nervous system it affects.
b What will people with this disease have difficulty doing?

4 People taking part in motor sports wear head, neck and back protection. Why is it important to protect each of these areas?

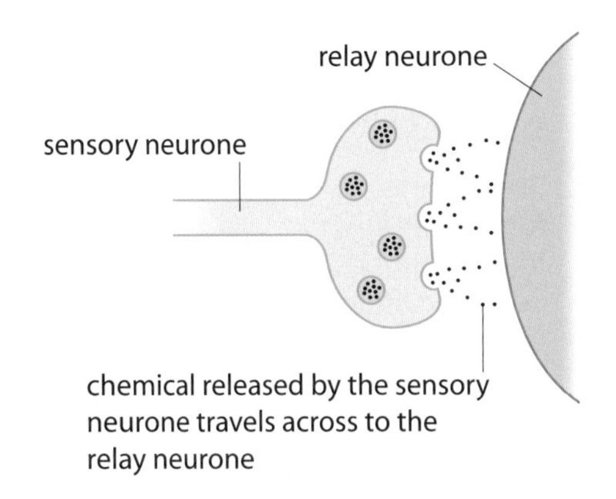

C What happens at a synapse.

5 What part of the body is not used in a spinal reflex action?

6 Coughing, sneezing and blinking are examples of reflex actions. Identify what part of the body each reflex action protects.

7 Why does our body have reflex actions?

8 Describe the reflex action of somebody who accidentally pricks their finger on something sharp.

9 **a** Explain in detail how an athlete reacts when the gun is fired.
b Suggest how a sprinter might improve his reaction time?

P P D P P

Drugs and their effects on the CNS

By the end of this topic you should be able to:

- describe different types of drugs and why some people use some of them for recreation
- explain that drugs change the chemical processes in the body and that in some cases this can lead to addiction
- describe how alcohol affects the nervous system and the effects too much alcohol can have on the body.

A **drug** is any chemical that alters how our body works.

Drugs that affect our **central nervous system (CNS)** control the movement of chemicals across the synapse. The natural chemicals in our nervous system have shapes that fit like a key in a lock. Drugs have similar shapes to these chemicals and mimic (copy) what they do.

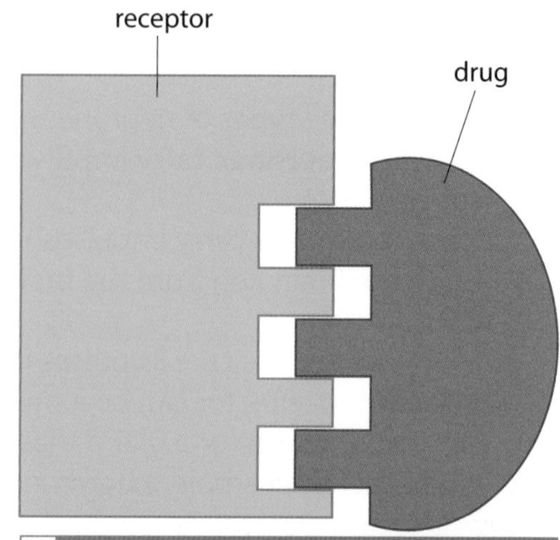

A Drugs mimic the chemicals released across the synapse.

Drug type	Effect	How they work
Stimulants for example amphetamines, ecstasy, caffeine, cocaine	make you more alert; stop you feeling tired; can make you more competitive and aggressive	speed up the release of chemicals across the synapse
Sedatives for example tranquillisers, barbiturates, alcohol, nicotine (tobacco), cannabis and anaesthetics	help you to relax and calm down	slow down the release of chemicals across the synapse
Analgesics for example painkillers, aspirin, opiates, including heroin and morphine	block pain receptors in the brain	block the release of chemicals across the synapse
Hallucinogens for example solvents, LSD, cannabis	make you see things differently – not as they really are; confuse your mind	interfere with the chemical messages

B Drugs that affect the central nervous system.

Drugs are not just taken for medical reasons; some drugs are also taken for pleasure. These drugs are called recreational drugs. Some recreational drugs are legal (for example, alcohol, caffeine and tobacco) but other recreational drugs are illegal (for example, cannabis, cocaine and ecstasy).

1 If you have a headache or toothache, you might take paracetamol or aspirin.
a Why do you take the drug?
b What type of drug are they?
c How do they stop the aching?

2 In the Second World War some pilots were given amphetamines.
a Suggest why they were given the drug.
b How did the drug help them fly better?

3 Which type of drug might a doctor give to somebody who is:
a always tired
b suffering from stress
c in pain?

4 Tea and coffee contain caffeine.
a Why do people like to drink tea and coffee during the day?
b How does the drink help them?

5 a Why do some people smoke tobacco when they are feeling stressed?
b What type of drug is found in tobacco?
c How does it help reduce feelings of stress?

Alcohol is a very popular recreational drug.

Quantity	Effects
Small amounts	Helps you to relax, lowers tension and removes inhibitions, like shyness. Even in small amounts it can reduce your concentration and slow down your reaction time.
Larger amounts	Changes your emotions, making you moody or aggressive; makes your speech slurred and makes you feel drowsy.
Very large amounts	Makes you vomit, causes breathing difficulties and can even lead to unconsciousness or coma.

C

Long-term damage by alcohol includes various types of brain damage and **cirrhosis** of the liver. This limits blood flow to the liver and stops it from working properly.

The effects of alcohol

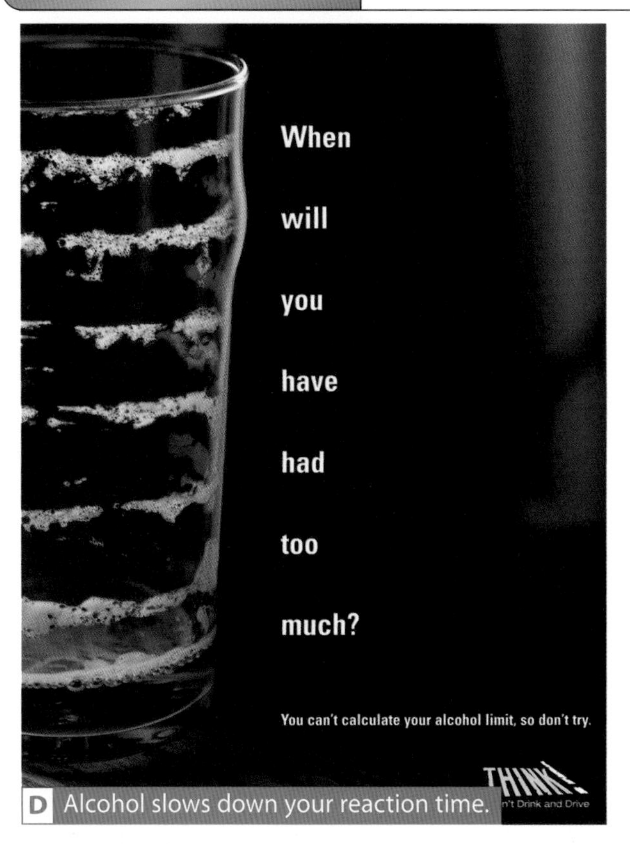

D Alcohol slows down your reaction time.

6 Why is alcohol such a popular recreational drug?

7 Why is it important not to drink and drive?

8 Which other drugs might affect your ability to drive a car properly?

9 Binge drinking is when you drink large amounts of alcohol in a short space of time. Suggest why binge drinking is so dangerous.

10 Drugs can also affect the performance of an athlete. Some cold and flu remedies contain analgesics, and stimulants or sedatives. A few athletes have been banned because traces of banned drugs have been found in their urine or blood sample. Some of these athletes have said that the only drug they have taken is to relieve cold or flu symptoms.

a Explain how each of these types of drug might help improve an athlete's performance.

b Explain why athletics committees have had to ban these drugs.

P D P P

Smoking and tobacco

By the end of this topic you should be able to:

- explain that tobacco smoke contains carcinogens and how the link between tobacco and cancer became accepted
- recall that nicotine is the addictive substance in tobacco
- describe how carbon monoxide in tobacco reduces the amount of oxygen the blood can carry
- describe how smoking in pregnancy can lead to a baby with low birth mass
- evaluate different ways to stop smoking.

It is more addictive than heroin. Its withdrawal symptoms are more severe than cocaine. It is responsible for more deaths than most other drugs put together. This drug is legal and is found in all types of tobacco.

The drug is called nicotine. If it is so dangerous then why has it not been banned?

Concerns about the effects of smoking on health were first raised around the year 1858. At about the same time smoking tobacco became a part of everyday life. The number of people smoking cigarettes grew until 1945.

Once inhaled, nicotine attaches to smoke and tar and is carried into the lungs. It is absorbed into the bloodstream and reaches the brain within 10 seconds. Your body quickly becomes tolerant of the nicotine, so you need to smoke more to get the same 'kick'. This leads to addiction.

1 What do you think happened as more people began to smoke?

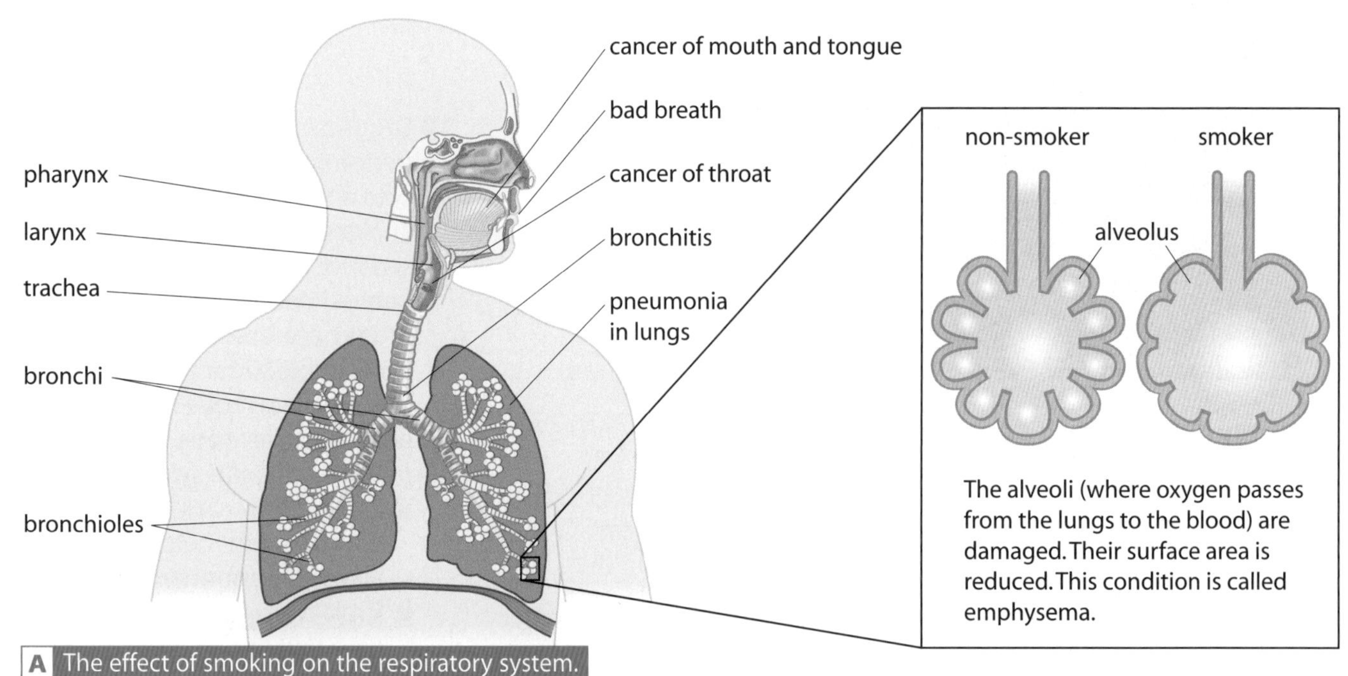

A The effect of smoking on the respiratory system.

The first evidence of a link between smoking and lung cancer was published in 1950. In 1964 the US Surgeon General announced that smoking tobacco causes lung cancer. It was not until 1998 that tobacco companies admitted that nicotine is addictive and that smoking may cause cancer.

It is now known that as you inhale the tobacco smoke, tar and nicotine are taken into your mouth. As the cigarette burns, some of the chemicals in it are changed into cancer-causing chemicals called **carcinogens**. They can cause cancers of the mouth, nose and digestive system, as well as lung cancer. It is estimated that smoking causes 30% of all cancer deaths.

Cigarette smoke contains a gas called **carbon monoxide**, which is absorbed into your bloodstream. Carbon monoxide attaches to the red blood cells and stops them from picking up oxygen.

Smoking during pregnancy means that the fetus cannot get the good supply of oxygen needed to grow properly. This increases the risks of miscarriage, premature birth and the baby being born underweight or even stillborn.

Many people who smoke tobacco know that it is dangerous, and want to give up. People try:

- just stopping
- hypnotism or acupuncture to stop the craving for nicotine
- replacing tobacco with herbal cigarettes (they still inhale tar but no nicotine)
- nicotine patches on the skin, which give a reduced amount of nicotine (the idea is that they gradually reduce the amount of nicotine they take in)
- gradually reducing the amount of cigarettes they smoke.

2 Why do people who smoke tire faster than if they did not smoke?

3 Which chemical in cigarette smoke makes people feel tired?

4 To improve stamina, athletes need a good supply of oxygen. How does smoking reduce a person's stamina?

5 **a** How does smoking affect an unborn child?
b How does this happen?

6 Make a list of how people might give up smoking.

7 Which methods are likely to be:
a most successful
b least successful?
Give reasons for each of your answers.

Measuring fitness

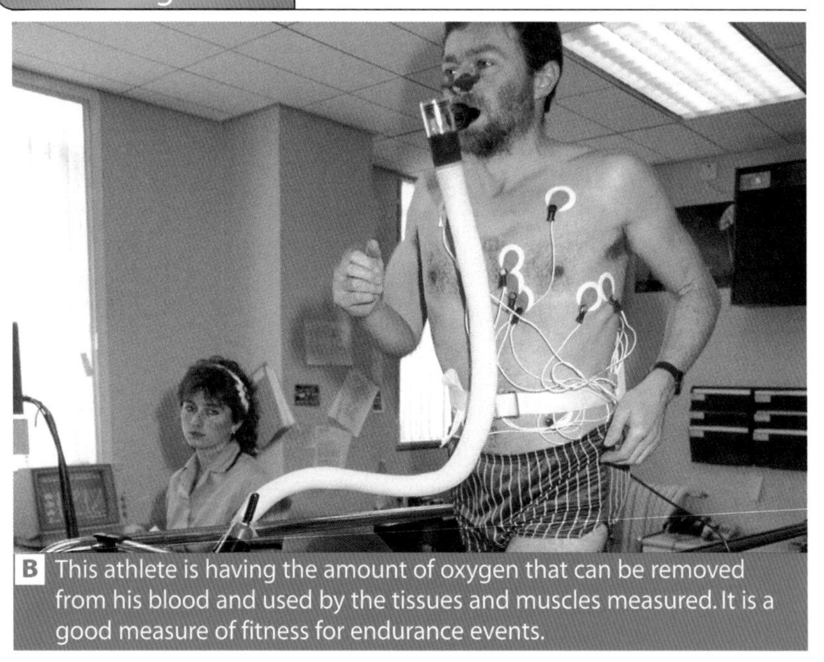

B This athlete is having the amount of oxygen that can be removed from his blood and used by the tissues and muscles measured. It is a good measure of fitness for endurance events.

8 Stamina is your ability to exercise for long periods of time. A person with lots of stamina needs to maintain a good supply of oxygen to the muscles.
a How will smoking tobacco affect an athlete's stamina?
b Explain your answer to part **a**.

Drug abuse

By the end of this topic you should be able to:

- describe different types of drugs and their effect on the human body
- evaluate claims about the effect of cannabis on health and its link to hard drugs
- explain that some drugs are legal and some are illegal
- describe the overall impact of legal and illegal drugs on health.

People take drugs for recreational or medical reasons. Most drugs were originally used to deal with injury or sickness. **Drug abuse** is when people take too much of a drug or use it for the wrong reasons. Spanish explorers used to chew on coca plant leaves to keep awake. Today it is used to make cocaine.

If some drugs are used a lot, your body builds up a **tolerance** to them. This means you must use more of the drug to get the same effect. As a drug is used more often and in greater amounts your body becomes more **dependent** on it. This means you will find it difficult to manage without the drug and will need to take it regularly. This leads to **addiction**. You are addicted when you cannot manage without taking the drug.

When you try to stop taking a drug you are addicted to, you suffer from **withdrawal symptoms**. These can include feeling sick, headaches and flu-like symptoms. More severe withdrawal symptoms include tremors and fits.

Alcohol DANGER OF DEATH

Also known as: booze

Taken by: drinking

Effects
- *Short term:* relaxation, slurred speech, drowsiness, reaction time reduced, poor concentration
- *Long term:* vitamin deficiency, addiction, memory loss, reduced brain function, liver damage

Withdrawal symptoms: anxiety, nausea, delirium tremens (DTs, shaking), insomnia, hallucinations, fits (seizures)

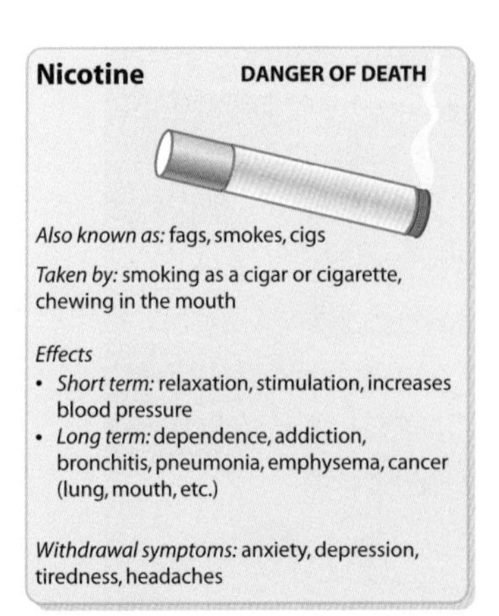

Nicotine DANGER OF DEATH

Also known as: fags, smokes, cigs

Taken by: smoking as a cigar or cigarette, chewing in the mouth

Effects
- *Short term:* relaxation, stimulation, increases blood pressure
- *Long term:* dependence, addiction, bronchitis, pneumonia, emphysema, cancer (lung, mouth, etc.)

Withdrawal symptoms: anxiety, depression, tiredness, headaches

Cocaine: DANGER OF DEATH

Taken by: snorting as a powder, smoking, injecting

Effects
- *Short term:* feeling of excitement, euphoria (rush of feeling good), reduced hunger, feeling strong, high lasts about 1 hour folllowed by a period of depression
- *Long term:* addictive, insomnia, depression (sad and moody), dizziness and headache, anxiety (worried and nervous), movement problems, hallucinations, too much cocaine (overdose) can kill you.

Withdrawal symptoms: depression, anxiety, paranoia, exhaustion

Cannabis:

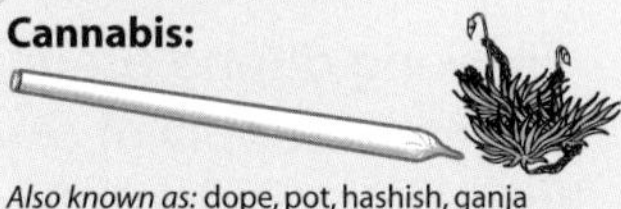

Also known as: dope, pot, hashish, ganja

Taken by: smoking, eating as cake

Effects
- *Short term:* relaxation, poor co-ordination, reduced blood pressure, sleepiness, lack of concentration
- *Long term:* **psychological dependence** (thinking that you need the drug), hallucinations, memory loss, delusions (seeing things that are not really happening), smoking-related problems (bronchitis, emphysema, lung cancer)

Heroin DANGER OF DEATH

Also known as: smack, crank, horse

Taken by: injecting, smoking, snorting

Effects
- *Short term:* reduced pain (analgesia), euphoria, nausea (feeling sick), hypothermia (severe drop in internal body temperature), breathing difficulties, drowsiness
- *Long term:* danger of death due to overdose, collapsed veins, infections, HIV/AIDS, liver damage (hepatitis)

Withdrawal symptoms: anxiety, fever, muscle cramps, diarrhoea, chills

Caffeine

Also known as: tea, coffee, cola, chocolate

Taken by: drinking mostly

Effects
- *Short term:* more alert, insomnia, headaches
- *Long term:* dependence, addiction, insomnia

Withdrawal symptoms: headache, fatigue, muscle pains

A Some recreational drugs and their effects.

Drug	Dependence	Withdrawal	Tolerance	Intoxication	Death rate
Nicotine	6	4	5	2	6
Heroin	5	5	6	5	4
Cocaine	4	3	3	4	3
Alcohol	3	6	4	6	5
Caffeine	2	2	2	1	1
Cannabis	1	1	1	3	2

B The effects of a number of drugs compared with each other (1= lowest; 6=highest).

1 Explain which of the drugs in table B your body becomes:
   a most easily tolerant to
   b most dependent on
   c most addicted to?

2 What are withdrawal symptoms?

3 'Tonics' sold to people in America in the 1800s contained heroin.
   a What would a person feel having taken this 'tonic'?
   b What long-term problems might they have suffered?

4 Why might taking heroin lead to you getting HIV, AIDS or other infections?

Some drugs are illegal for people to buy, sell or take because they are dangerous. The government has placed drugs into groups according to how dangerous they are.

Classification	Drugs included
A	heroin, LSD, opium, cocaine, crack, ecstasy, processed magic mushrooms
B	amphetamines, barbiturates, codeine, DF118 painkillers
C	mild amphetamines, tranquillisers, cannabis

C Class A drugs are sometimes called 'hard drugs', and class B and C 'soft drugs'.

Cannabis is described as a **'gateway drug'** because it is thought that people who use it will go on to take harder drugs like heroin and cocaine. Research has shown that most users of class A drugs have used cannabis. They also found that most cannabis users do not go on to use other hard drugs regularly.

Some illegal drugs also can have medical benefits. For example, the active chemical in cannabis is used to reduce the symptoms of glaucoma (high pressure in the eye) and multiple sclerosis (a disease that affects the nervous system).

5 Why are heroin and cocaine described as 'hard' drugs?

6 Why do more people die from using nicotine and alcohol than heroin and cocaine?

7 **a** How dangerous does cannabis seem to be compared with other drugs?
   b Cannabis is often mixed with tobacco and smoked. How would this affect the health of the cannabis user?
   c Why is cannabis said to be a 'gateway drug'?
   d Why are the effects of long-term cannabis use only just being noticed?

P

P

D

P

P

Hormones

By the end of this topic you should be able to:

- describe how some processes in the body are controlled by hormones
- describe how hormones are secreted by glands and are carried by the bloodstream
- explain how hormones regulate many of the functions of several organs and cells.

P

V

A

A The hormone adrenaline is released in times of excitement and stress.

Hormones are chemical 'messengers'. They are released (secreted) from **glands** around the body. They travel through the bloodstream to organs where they control important processes (such as growth and puberty). Unlike nerve impulses, the movement of hormones is slower and their effects longer term. The types and amounts of hormone released depend on the needs of the body at a particular time. For example hormones that control sexual development are mainly released at certain times in our lives, such as during puberty.

Pituitary gland secretes 'stimulating' hormones that activate other glands. It also releases a hormone (**anti-diuretic hormone**) that controls water levels in the blood, and **growth hormones** that control the size of the body, the length of limbs and muscle growth. Growth hormones also help the body to recover after injury.

Thyroid gland secretes a hormone which controls **metabolism** (energy levels). Low levels of the hormone (thyroxine) can make you feel tired and slow, high levels make you overactive.

Pancreas secretes hormones which control blood sugar (glucose) levels. The hormone insulin lowers blood glucose levels and the hormone glucagon raises glucose levels.

Adrenal gland secretes adrenaline which prepares the body for action. It is called the 'fight or flight' hormone. The adrenal gland is stimulated by the nervous system. When it is released it makes your heart beat faster. It diverts blood from the digestive system and skin and sends it to the muscles.

B Some examples of glands and hormones.

1 Produce a table showing the major glands, the hormone they release and the job the hormone does.

Blood passing through the brain is carefully monitored for many things including changes in temperature, glucose concentration and water. The brain sends hormones to the pituitary gland which then sends hormones into the bloodstream. These hormones travel to other glands in order to get them to secrete the hormones needed.

2 When you watch a scary film you might go pale, your heart might beat faster and your muscles may tighten. Which hormone causes this?

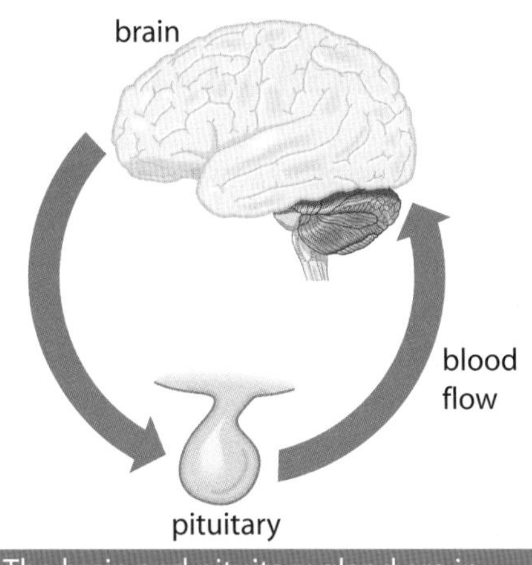

C The brain and pituitary gland are in constant communication.

D Ready for action.

3 Athletes perform best when they are 'pumped up' and ready for action.
 a What hormone causes this?
 b How does it work on the body?
 c How does this help improve an athlete's performance?

4 Some athletes take growth hormones illegally.
 a Suggest why they do this.
 b Why do you think growth hormones are banned by sporting organisations?

5 Some farmers inject their cattle, sheep, pigs and chickens with growth hormone. Suggest why they do this.

6 Suggest how you would be affected if you had an overactive thyroid gland.

7 a For each of the following, name the hormone that controls the reaction in your body and the gland that secretes the hormone.
 (i) sugar levels in the blood
 (ii) energy and metabolism
 (iii) readiness for action
 (iv) male sexual development
 (v) female sexual development.
 b Which gland controls all the other glands in your body?

Controlling our internal environment

A How does the athlete keep cool?

By the end of this topic you should be able to:

- describe how body temperature is kept at the level at which enzymes work best
- explain how blood sugar provides cells with energy
- describe how water leaves the body via the lungs, skin and kidneys
- describe how ions are lost through the skin as sweat and the kidneys as urine.

When you exercise you get hot, go red, sweat and breathe faster and deeper. The hairs on your arms lie flat and skin pores open. This happens because your body is trying to stay cool.

When you get cold the opposite happens. Your skin goes pale, pores close up and hairs stand on end. You might even start to shiver. Your body is trying to keep warm. If your internal body temperature gets too low you may suffer from hypothermia.

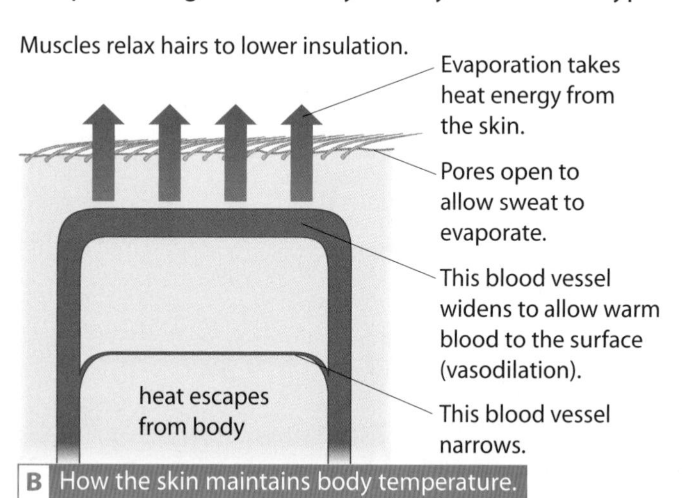

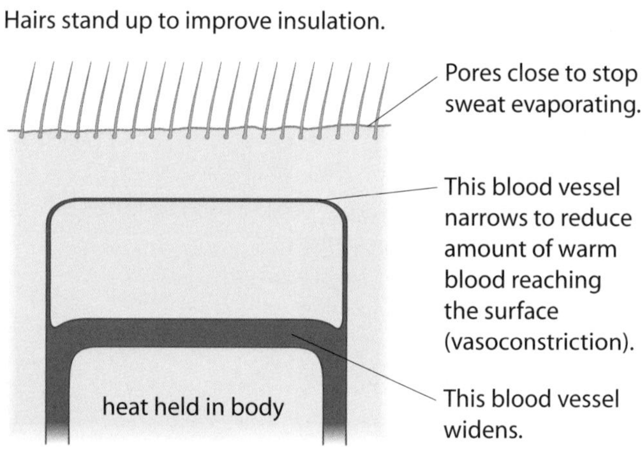

B How the skin maintains body temperature.

1 Long-distance swimmers risk getting hypothermia.
  a Suggest why they cover themselves with Vaseline.
  b Why might hypothermia cause them to drown?

2 Why do you go red when you exercise?

3 When people take drugs like ecstasy they find it difficult to cool down. Suggest what is happening to their body.

Your body tries to stay at a steady internal temperature, around 37 °C. This is the temperature at which **enzymes** in your body work best. Enzymes speed up chemical reactions in the body. Without enzymes your body would not be able to work properly. The process of keeping things constant and balanced in your body is called homeostasis.

Homeostasis uses **negative feedback** to control conditions in your body. Negative feedback checks the levels and acts to reduce them if they start to rise or increase them if they start to fall.

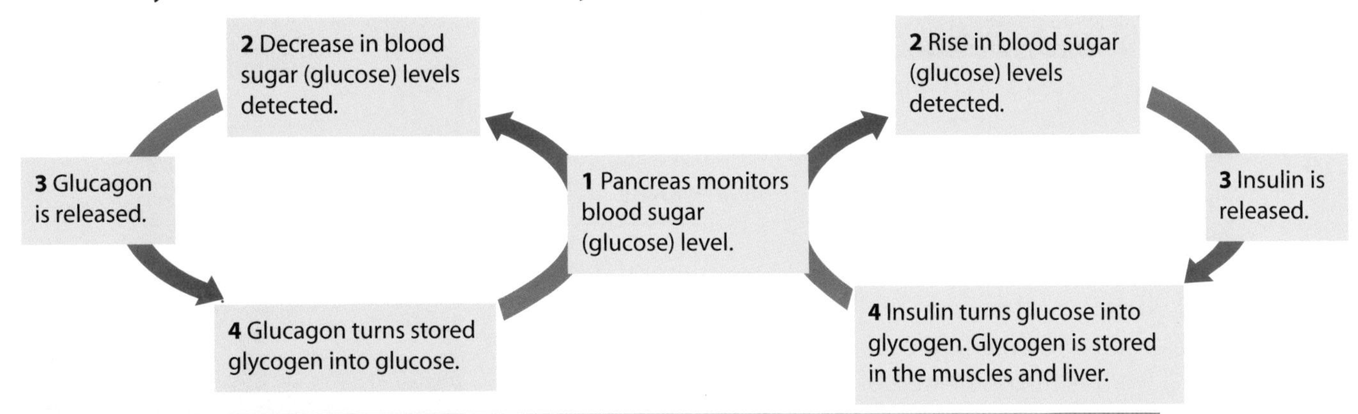

C How the body controls blood glucose levels. An example of how you control conditions in your body.

Blood sugar (glucose) concentration is an example of homeostasis. Glucose is a source of energy for cells. The amount of glucose in the blood is constantly monitored by the pancreas.

If too much sugar builds up in your blood you will get tired and thirsty. If left untreated it can lead to weight loss, unconsciousness and death.

Energy drinks

4 Companies advertise some drinks as being better than water at 'replacing lost energy' if you exercise.
- **a** How could you find out if their claim is correct?
- **b** What would the variables be?
- **c** Which variables would stay the same?
- **d** Which variable would you change?

You take in water when you eat and drink. You lose water when you sweat, breathe out and urinate.

Your kidneys control the amount of water in your blood. If the amount of water in the blood is too low a hormone called ADH (anti-diuretic hormone) is released from the pituitary gland. ADH activates the kidneys to excrete less water from the blood. Water that is filtered out in the kidneys is returned to the bloodstream to stop you from losing too much water (**dehydrating**).

5 Draw a flow chart similar to diagram C to show the control of water in the bloodstream.

Salt (sodium chloride) contains sodium and chloride **ions**. These are needed to help our body work properly. Too much salt can be dangerous, but so is too little. Sodium and chloride levels in the blood are controlled by the kidneys. These ions are also lost when we sweat.

6 **a** If you sweat a lot what will happen to your:
- **i)** salt levels
- **ii)** water levels?

b Why might this be dangerous?

7 Many drinks cause the body to lose water. They are called diuretics. Alcohol, caffeine and fizzy drinks are all diuretics. Why is it not a good idea to drink these when you feel thirsty?

8 Look at photograph A. The liquid she is drinking is isotonic. This means it is the same concentration as body fluids.
- **a** How will this help her maintain a balance?
- **b** What would happen if she did not take in these fluids?

V P V P P D P P

Menstruation

By the end of this topic you should be able to:

- explain the role of FSH, LH and oestrogen in the menstrual cycle
- explain the role of FSH as a fertility drug
- describe the role of hormones in oral contraceptives.

Following puberty, women have a period once a month. It is called **menstruation**. Menstruation is when the lining of the uterus is released and bleeding occurs from the vagina for about five days. It is the end result of the **menstrual cycle**. This is controlled by hormones (diagram C).

Just after menstruation an immature egg starts to develop in an ovary. As it matures a hormone (**follicle stimulating hormone** (**FSH**)) is released from the pituitary gland. FSH makes the ovary produce a protective covering (follicle) around the egg.

Just before ovulation the pituitary gland secretes a hormone (luteinising hormone). This triggers ovulation and turns the follicle into a yellow body.

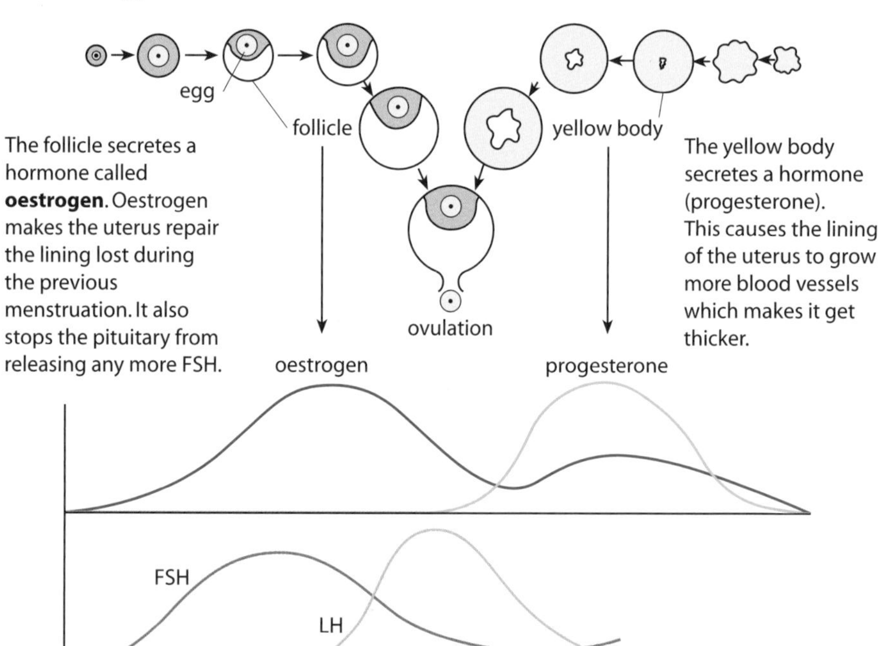

The follicle secretes a hormone called **oestrogen**. Oestrogen makes the uterus repair the lining lost during the previous menstruation. It also stops the pituitary from releasing any more FSH.

The yellow body secretes a hormone (progesterone). This causes the lining of the uterus to grow more blood vessels which makes it get thicker.

After about two weeks the egg is released into the oviduct (egg tube) this is called ovulation. The egg begins its journey to the uterus.

menstruation

0 2 4 6 8 10 12 14 16 18 20 22 24 26 28

Days

C Hormones control the menstrual cycle.

A To have the best chance of winning in sport you need to be at peak fitness. Where an athlete is in her menstrual cycle can affect her performance.

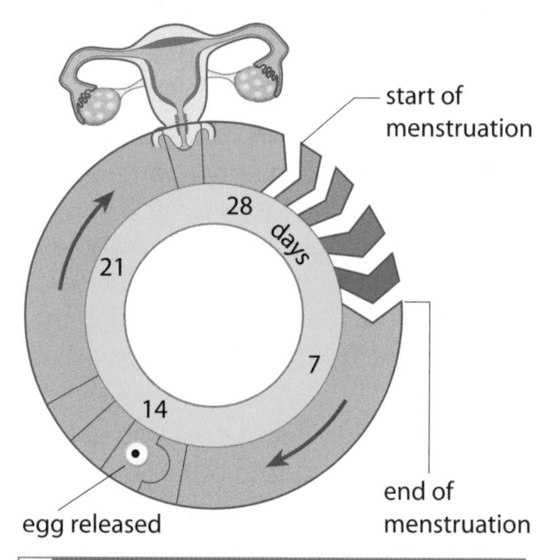

B Menstruation is one part of the menstrual cycle.

1 Look at diagram C. Complete table D for each of the hormones involved in the menstrual cycle.

Hormone	Where it is secreted from	Where it travels to and what it does when it arrives

D

2 How many days are there in a typical menstrual cycle?

3 Between which days is oestrogen at its highest levels?

If the egg is not fertilised, oestrogen and progesterone stop being secreted. This makes the yellow body wither away and the lining of the uterus is lost. Menstruation begins again.

If the egg is fertilised the yellow body stays in the ovary and continues to release progesterone. Oestrogen also continues to be secreted. This stops menstruation and no more eggs are released. The uterus lining gets thicker with more blood vessels. As the fertilised egg embeds itself in the wall of the uterus, the placenta begins to develop. The placenta releases oestrogen and progesterone instead. A sudden drop in oestrogen and progesterone starts the birth process.

Contraceptive pills contain the hormones oestrogen and progesterone. This stops the eggs from being matured by the ovaries and so fertilisation cannot happen.

Women that have difficulty becoming pregnant might be given a fertility drug to stimulate the production of FSH. This in turn makes the ovaries release an egg.

4 What happens to oestrogen levels if:
 a an egg is not fertilised
 b an egg is fertilised?

5 Explain how the contraceptive pill works.

6 Women that have fertility treatment often have multiple births (several children at the same time). Why might this happen?

7 The most difficult time for a female athlete to race well is the week before menstruation and the week after ovulation. The best time for female athletes to race is thought to be just before ovulation, between days 9 to 12.
 a Study the changing levels of oestrogen and progesterone in diagram C and explain the connection between the two hormones and training.
 b Design a training schedule for a female athlete so that she can train and race at peak performance. Think about the menstrual cycle and how it affects performance.
 c How might taking the pill help with race training and performance?

Food and health

By the end of this topic you should be able to:

- describe what is meant by malnourishment
- evaluate the effect of food on health
- describe the health problems of some people in the developing world because of lack of food
- explain how too much salt in the diet can lead to increased blood pressure
- recall that processed food often contains a high amount of fat and salt.

It may not be obvious just by looking at the people in the photographs in A, but they are both suffering from **malnutrition**. Malnutrition happens when you eat the wrong amount of each type of **nutrient** (too much or too little). Nutrients are the chemicals that are contained in food.

Vitamins play a very important part in almost all bodily processes, including the immune, nervous and hormonal systems. There are many different vitamins and minerals so it is important to eat a wide a variety of foods. Some vitamins are lost in cooking. Some vitamins are lost in urine.

Minerals are chemical elements such as calcium, sodium and potassium. You need more of some than others. Those you need more of are called 'major minerals', those you need less of are called 'trace minerals'. Minerals are needed by teeth and bones. Some help your heart to beat properly, and some help your nervous system and body cells to work efficiently. Others form part of enzymes and hormones.

Fats are used as energy stores in the body and contain important vitamins. They contain more energy than carbohydrates. Too much saturated fat in your diet is not good for you. Any fat that is not used up is stored in fat cells around the body. This makes you put on weight. **Unsaturated fats** mainly come from plants (e.g. sunflower oil, olive oil) and fish (as omega 3 oils). Unsaturated fats are much better for you than saturated fats as they contain many essential vitamins and minerals.

Water (essential for staying alive, 60% of your body is water).

Proteins are used to build muscles, to repair damaged tissue and to help young bodies grow. They are also used to produce important chemicals such as enzymes and hormones.

Carbohydrates give you energy. Examples include sugars and starches. Sugars such as glucose can be used straight away by the body. Starches are more difficult to digest. Starch has to be turned into glucose before it can be used.

B What is a balanced meal?

Fibre is important for a healthy digestive system. Cellulose and fibre are also carbohydrates. Cellulose is difficult for us to digest. It is often called fibre or roughage.

A What do the diets of these two people have in common?

1 The child in photograph A eats mainly one type of carbohydrate. Her food also contains a few vitamins and minerals. What else is missing from her diet?

2 The man mainly eats foods that are high in carbohydrate, protein, fats, salt and sugar. What is missing from his diet?

Malnutrition is an example of an unbalanced diet. A **balanced diet** is one that has the correct amount of each type of nutrient. An unbalanced diet can have too much or too little of one type of nutrient or energy.

Processed foods are foods that have been altered in some way. They are sometimes called convenience or fast foods because they don't take long to prepare. Although processed foods are made from fresh ingredients, other substances might have been added to improve their taste or colour, or to stop them from going bad. Salt or sugar are often added to improve flavour or act as a preservative. Saturated fats also help to improve taste and flavour.

3 The man in photograph A eats lots of processed food. What nutrients will he eat too much of?

4 If saturated fats are high in energy why are they not good for us in large amounts?

P

5 How can you tell that the child in photograph A has a diet that lacks protein?

C Why do you need vitamins and minerals?

V

Vitamins and minerals cannot be made in your body so you have to get them from the food that you eat. Because there are so many vitamins and minerals it is important to eat a wide variety of food. Vitamins and minerals often work together in your body. For example vitamin C helps the immune system, vitamin D and calcium are needed for healthy bones, iron is needed for oxygen uptake in the blood. Sodium and potassium are also needed to keep the nervous system working.

You can survive for a long time without food, but you can only last for a few days without water.

6 Why might drinking too much water not be good for you?

7 Why is it better to eat raw vegetables and fresh fruit rather than cooked or tinned vegetables and fruit?

8 It is important for athletes to have balanced diets.

P

a List all the ingredients of a balanced diet.

D

b Explain why each ingredient is important to an athlete.

P

c How might an athlete's diet be different from somebody who sits behind a desk all day?

P

Metabolism

By the end of this topic you should be able to:

- describe how the metabolic rate depends on the amount of activity you do and the amount of muscle and fat that you have
- explain how the less exercise you take and the warmer it is, the less food you need
- describe how people who exercise regularly are usually fitter
- explain how if you exercise your metabolic rate stays high for a short time after.

To live you need energy. The more exercise you do the more energy you need. Many chemical reactions take place in your body. These reactions release the energy from food. Your **metabolic rate** is the speed at which your body uses that energy.

A Does your work affect your metabolic rate?

Even when you are resting, you need energy to keep your heart beating, and for breathing and digestion. This is called your resting or basal metabolic rate.

The higher your metabolic rate the more energy you use. The more energy you use the less fat you store in your body. The less energy you use the lower your metabolic rate and the more fat you store in your body.

1 Look at the photographs in A. Who is likely to have the higher metabolic rate? Explain your answer.

2 How does your basal metabolic rate change if you are ill with a fever? Explain why it changes.

As you exercise your body builds up muscle tissue. This increases your metabolic rate. As you exercise your body has to work harder and so more energy is needed which increases your metabolic rate further. Your body uses some of its fat stores to replace the energy lost. If you exercise regularly you use energy more efficiently. This makes your body fitter and leaner as your muscles tone up and excess fat is used. Even when you have finished exercising your metabolic rate will remain high for a little while after.

If you do not eat, your body loses muscle tissue and your energy levels drop. This makes your metabolic rate fall and prevents your body from running out of energy completely. If it does you die! People who are suffering from malnutrition have a lower metabolic rate.

3 Explain why dieting on its own is not the best way to lose weight.

4 Why is exercising important if you want to lose weight?

B Different forms of exercise affect your metabolic rate differently.

When you exercise your metabolic rate increases. As this happens your body uses up some of its energy that is stored as glycogen in the muscles. Eating carbohydrates, such as glucose, helps replace the lost energy and maintains a high metabolic rate. If you do not exercise much then your glycogen stores will not need to be replaced as often, and so you can eat less. Similarly, on a warm day it doesn't take so much energy to keep warm, so you can eat less.

5 Which of the activities in B is most likely to increase the athlete's metabolic rate?

Everybody has a different metabolic rate. It depends on many factors including inheritance and how you live your life. Your metabolic rate gets less with age.

6 Explain why your metabolic rate might decrease as you get older.

7 What advice would you give to people as they get older to help them to maintain a high metabolic rate?

8 **a** Make a list of things that affect your metabolic rate.

b If your metabolic rate is low how can you increase it?

c What is the connection between metabolism and general fitness?

B1a.10

Cholesterol

By the end of this topic you should be able to:

- describe how the amount of cholesterol produced by the liver depends on diet and inherited factors
- describe how high levels of cholesterol in the blood lead to an increased risk of heart disease
- explain how cholesterol is carried around the body by two different types of lipoprotein and how the balance of these is important to good heart health
- explain how the level of cholesterol in the blood is influenced by the amount and type of fat in the diet
- describe the effect of statins on cardiovascular disease.

You start to get fat when you use less energy than you take in from the food you eat. Extra glucose is turned into fat and stored in fat cells under the skin. You don't gain fat cells – they just get bigger as they fill up with fat.

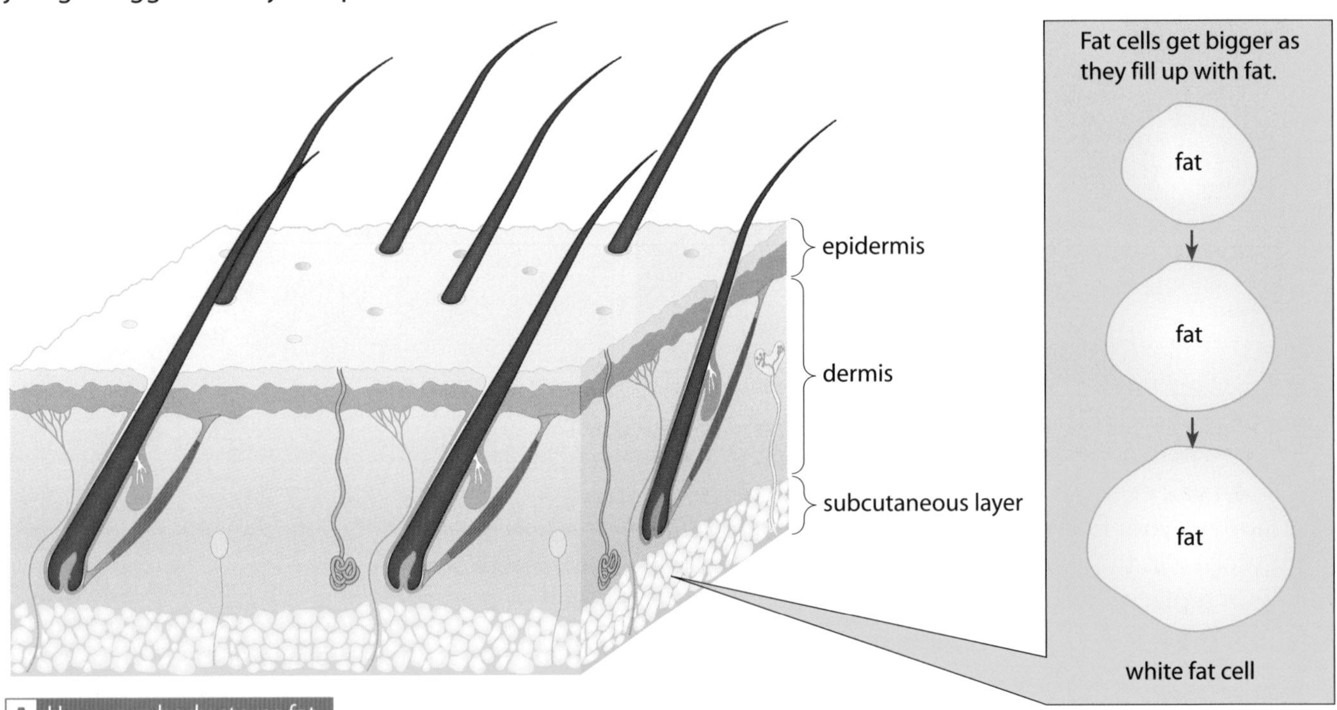

A How your body stores fat.

Your body needs some fat in order to work properly. This is called **essential fat**. Overweight people have too much fat in their body for their size. There are over a thousand million people in the world who are overweight and 200 million who are **obese**.

1 In which parts of the world would you expect to find the most obese people? Explain your answer.

Cholesterol is a type of fat made by the liver. It is found in every cell in your body. It is carried around in your bloodstream attached to a protein. This fat–protein combination is called a lipoprotein. Lipoproteins can be high density (HDL) or low density (LDL).

LDL is mostly fat with only a little protein. It is called 'bad cholesterol' because it deposits fat in your arteries, making them narrower. This means that blood cannot pass through them easily. If they block up completely you might have a heart attack or a stroke. High cholesterol in the blood can be inherited but it is also caused by a diet that is high in **saturated fats** (e.g. dairy products, red meat, cakes and biscuits).

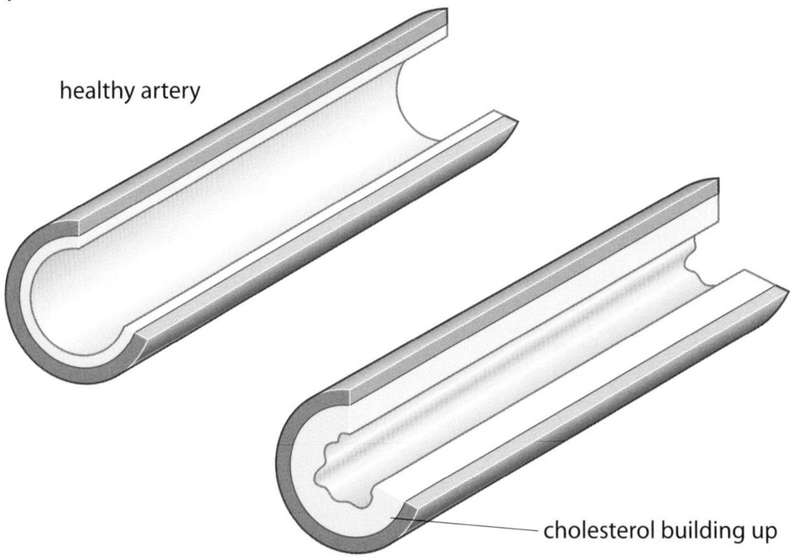

B High levels of LDL cholesterol clog your arteries.

HDL is mostly protein with only a little fat attached. It is called 'good cholesterol' because it helps to stop 'bad cholesterol' from building up in your arteries. The balance of good and bad cholesterol is very important. The amount of bad cholesterol in your blood can be reduced by eating polyunsaturated fats (e.g. vegetable oils, low-fat spreads and fish). Mono-unsaturated fats do not affect cholesterol in your blood.

Drugs can be taken to lower the amount of cholesterol in your blood. The main group of drugs are called **statins**. They reduce the amount of cholesterol made in the liver. Statins are usually given to people who have not been able to lower the amount of cholesterol in the blood by changing their diet. Statins can cause headaches, stomach pains, nausea and liver problems.

2 Make a list of foods that are:
  a high in saturated fats
  b low in saturated fats.

3 Why is it important to try and reduce your cholesterol by changing your diet before taking statins?

4 List the differences between HDL and LDL. Show these differences in a table.

Diet problems

By the end of this topic you should be able to:

- describe how some people in the developing world suffer from health problems linked to lack of food
- evaluate claims made by slimming programmes.

People usually become overweight or obese because they eat too many high-energy foods containing saturated fats or sugar. If they do not exercise enough these high-energy foods are stored as fat.

Being overweight or obese can lead to many health problems.

- Your body has to carry around more weight. This puts a strain on your heart and makes your blood pressure rise. If your blood pressure increases and your blood vessels get narrower as fat is deposited in them, the amount of oxygen reaching the heart is reduced. This means that you might suffer from heart disease. If blood circulation to the heart is stopped completely you suffer a heart attack.
- Your bones, joints and your muscles also have to support extra weight. In time the cartilage at your joints starts to wear and your bones rub together. The joints become sore and inflamed. This is called arthritis.
- Diabetes can be caused by being overweight. It occurs if you do not produce enough insulin to meet the needs of high sugar levels in your diet.

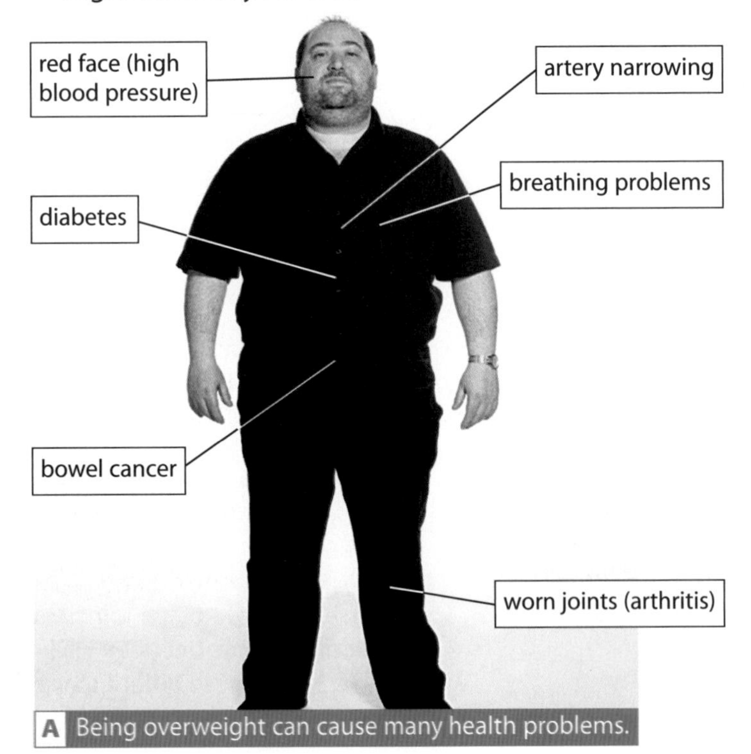

A Being overweight can cause many health problems.

When people are overweight they are advised to lose weight. They do this by going on a **diet**. Some contain less saturated fat, others fewer carbohydrates.

Atkins Diet

1st stage

- You only eat proteins (meat, fish, poultry, eggs) and fats (oils, butter, etc).
- You are only allowed 20 g of carbohydrate each day.
- You don't eat any fruit, vegetables or bread. This tricks your body into thinking it is starving and your body uses up its glycogen store.
- Your body loses a lot of water at first.
- If your body is starving it might begin to break down muscle tissue.
- It might also damage your kidneys.

2nd stage (after a couple of weeks)

- You start to eat more carbohydrate until you stop losing weight.
- This gives you your maximum carbohydrate limit.
- You must stay under that limit to carry on losing weight.

Cabbage Soup Diet

- Take a large white cabbage and slice it up. Put it in a pan and cover it with water. Boil it until it becomes a soft pulp. Season to taste.
- You can eat as much cabbage soup as you like.
- It is low in fat but high in fibre.
- You can combine the soup with any fruit or vegetable you like (except for corn, beans, peas and bananas).
- You only stick on this diet for 7 days then move onto a longer term dieting plan.

Weight Watchers

- Uses a points-based system.
- Each food is given points based on the amount of fat, fibre and energy it contains.
- Each person is set a points target for the day.
- You can eat anything as long as you do not go over the target.
- You can meet once a week with other 'weight watchers' to measure and discuss progress.

Slim Fast Diet

- Slim Fast is a meal-replacement diet.
- You can eat as often as six times a day to avoid highs and lows.
- You take two Slim Fast shakes for breakfast and for lunch.
- You have a normal dinner.
- Each Slim Fast meal replacement is around 240 calories and with the meal you should not go over 1200 calories.
- Slim Fast shakes contain added vitamins and minerals, essential fatty acids and proteins.

B There are many approaches to dieting.

1 Look at the diets shown in B.
 a Identify what is missing from each diet.
 b Describe briefly how each is meant to help you lose weight.
 c Is each diet balanced?

You can suffer serious health problems if you do not eat enough food.

- If you do not eat enough carbohydrate your energy levels drop and your metabolism slows down. You get tired easily and find it difficult to work.
- Lack of protein in your diet causes your muscles to waste (break down), making you weaker.
- If your body is starving it uses your muscles as a source of energy. You find it harder to move around and your metabolic rate falls.

An unbalanced diet might mean that you do not eat enough vitamins and minerals. These are important in fighting disease and infection. Some diseases, for example flu, can make you ill if you are normally healthy, but can kill you if your body cannot fight them.

If a woman's **lean body mass** falls below 20% then her periods may become irregular or might stop altogether.

2 Make a list of health problems people in the developing world might suffer from if they cannot get enough food to eat.

3 Why do starving people look so thin?

4 Why is it important that people in developing countries have enough food?

5 How are the dietary problems of the developed world:
 a different from the dietary problems of the developing world
 b similar to those in the developing world?

Microorganisms

By the end of this topic you should be able to:

- explain what causes infectious disease and how diseases are spread
- recall that disease-causing microorganisms are called pathogens
- describe the contribution made by Ignaz Semmelweiss to controlling infection in hospitals today.

Microorganisms are the tiny living things that can only be seen using a microscope. They are everywhere, including the food you eat and you!

Microorganisms

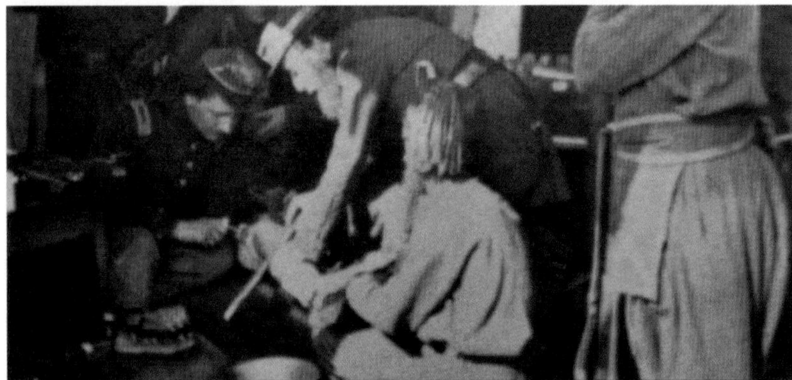

B Damaged limbs used to be amputated and the stump dipped in boiling tar. This was to stop the bleeding and prevent infection. Most people still died from their wounds as infection took hold.

Ignaz Semmelweiss was a doctor in the mid 1800s. He wondered why so many women died of 'childbed fever' soon after giving birth. He also noticed that student doctors carrying out work on dead bodies did not wash their hands before delivering a baby. When he got them to wash their hands in calcium chloride solution before delivering babies, fewer women died. He concluded that something was carried by the doctors from the dead bodies to the women.

Louis Pasteur and Joseph Lister studied what became called 'microorganisms'. Pasteur proved that there were '**germs**' in the air and that they carried infection and disease. Lister developed a special soap called carbolic soap. He insisted that all medical instruments, dressings and even surgeons should be cleaned with it before any operation. Lister's patients stayed healthy.

Chemicals that are used to clean wounds or get rid of sores are called **antiseptics**. Chemicals that are used to clean work surfaces and other places where pathogens might be found are called disinfectants.

Hygiene and hospitals

NHS bugs kill 5000 a year

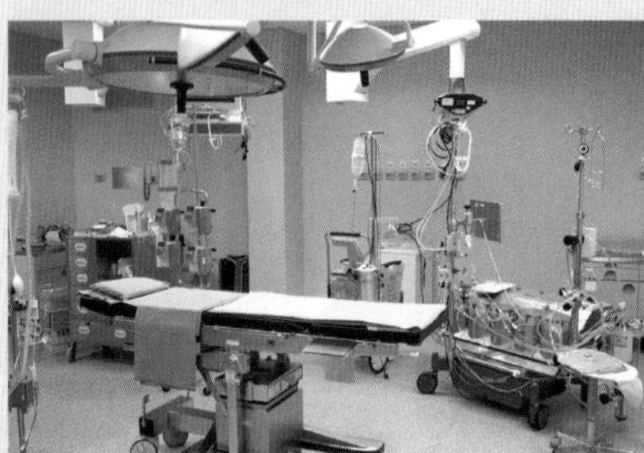

Up to 5000 people die each year from infections picked up in hospitals in England. The problem actually affects 100 000 people and costs the NHS a thousand million pounds. It is thought that deadly infections are spread because hygiene rules are broken. For example doctors and nurses not always washing their hands between treating patients A senior nursing officer said: 'Levels of cleanliness have deteriorated in recent years. I have seen dust under beds, cotton wool buds on the floor and dirty needles dumped in discarded meal trays. These guidelines are about changing the curtains around beds, cleaning floors and cleaning bathrooms'. Potentially fatal infections are carried in dust mites and a study has shown that improving ward cleanliness can reduce infections.

A Hospitals can make you sick.

1 Why is it important to keep hospitals clean?

2 What is:
a an antiseptic
b a disinfectant?

3 What antiseptic was used by Semmelweiss?

Microscopes

It was not until the 1870s that Robert Koch saw and identified 'germs'. He used a new and more powerful microscope to watch bacteria multiplying. He did experiments to find out how to stop them reproducing.

Microorganisms that cause illness or disease are types of **pathogens**. They are the 'germs' identified by Semmelweiss, Pasteur and Lister. There are four main types:

- **bacteria** (cause cholera, boils, MRSA, typhoid, tuberculosis, for example)
- **viruses** (cause warts, herpes, polio, flu, mumps, measles, smallpox, for example)
- fungi (cause athlete's foot, ringworm, for example)
- parasites (lice, fleas, 'worms', for example). (Note: these parasites are not microorganisms.)

D Microorganisms are passed from one person to another in many ways.

Keeping clean

Semmelweiss showed that keeping things clean helps to stop the spread of pathogens. We call this **hygiene**. It is about keeping things clean to reduce the risk of disease. Washing removes the dirt and grease that pathogens stick to and use as a source of energy to multiply.

E It's important to wash after exercise.

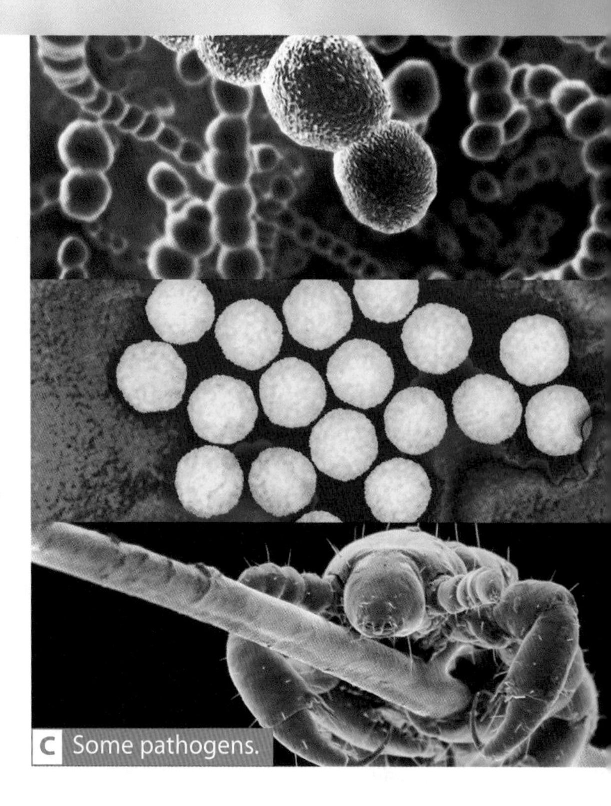

C Some pathogens.

4 Make a list of all the illnesses you can think of. Find out:
  a what type of pathogen they are
  b how they are transferred.

5 **a** Make a list of how many things you touch as you move from the toilet back into the room where you came from.
  b Why is it important to wash your hands after going to the toilet?

6 There are many new products that you can buy that are described as 'antibacterial'. What does this mean?

P

7 When you exercise you sweat a lot. Sweat is not just water, it also contains things like salt and proteins.
  a Why does sweat help bacteria multiply?
  b Where have the bacteria come from?
  c If you do not wash after exercising why does your body start to smell?
  d Why is it important to wash after exercising?

D P P

Protection against disease

By the end of this topic you should be able to:

- describe how your body defends itself against pathogens
- explain that bacteria and viruses spread rapidly and may produce toxins
- describe how viruses damage the cells in which they reproduce.

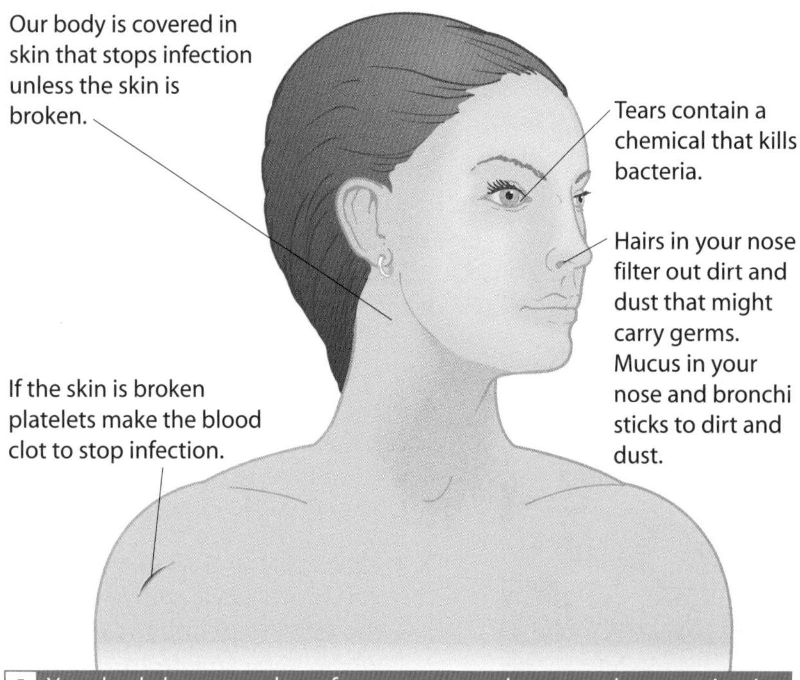

A Your body has a number of ways to stop micro organisms getting in.

Once pathogens get inside us they multiply. Your body's defence is **white blood cells**. There are two types: phagocytes and lymphocytes.

Some white blood cells (phagocytes) find a pathogen, surround it and then ingest it. This means swallowing it and breaking it down. Where there are lots of pathogens the area becomes red and swollen as these cells attack. This reaction is called **inflammation**. Dead white cells and pathogens collect to form pus.

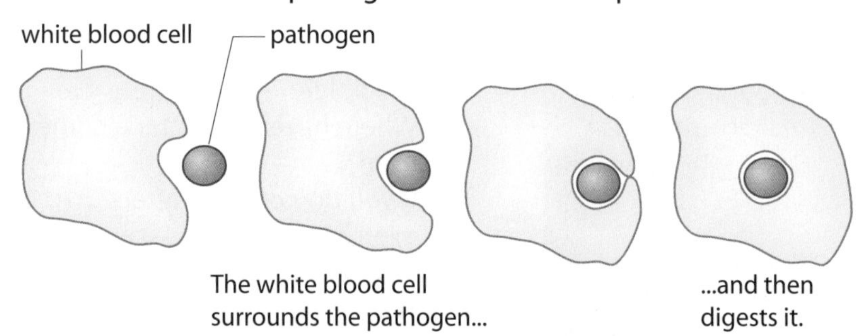

C White blood cells (phagocytes) ingest pathogens.

B Your body needs more oxygen when you exercise.

1. Why is it a good idea to breathe through your nose when exercising?
2. How might germs get underneath your skin?
3. Why is it important to wash regularly?
4. What do athletes do that makes them more likely to get coughs and colds?

Pathogens have molecules called antigens attached to them. These antigens alert some other white blood cells (lymphocytes). These make chemicals called **antibodies** that stick to antigens. Different white blood cells produce different antibodies. Once these antibodies have been made they stay in your body.

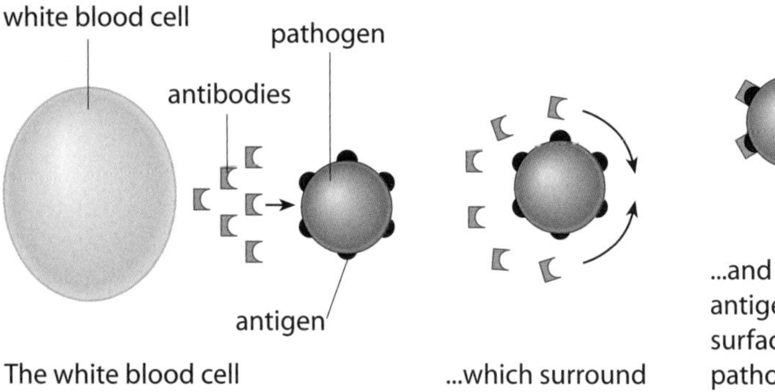

D Some white blood cells (lymphocytes) produce antibodies to attack a pathogen.

The pathogen can then be destroyed in a number of ways:

- The antibody makes the pathogen burst.
- The antibody sticks to the antigen and allows a white blood cell to destroy it.
- The antibodies cause the pathogens to stick together so that white blood cells can attack.

Bacteria and viruses make you ill by releasing poisonous chemicals called **toxins**. Toxins are destroyed by **antitoxins**.

5 If you cut yourself, germs from the soil may enter your bloodstream. List the ways in which your body can fight them.

Once they are inside you, bacteria can multiply rapidly – doubling in number about every 20 minutes. Sometimes your lymphocytes cannot make antibodies quickly enough and so a pathogen can make you ill. You might get symptoms such as a headache, fever or feeling sick.

Viruses multiply by entering the cells in your body. They use the chemicals inside the cell to make copies of themselves. These copies then travel around your body making more copies. Eventually, your body cannot cope with the number of viruses and you become ill.

Infectious diseases are spread because pathogens are passed from one person to another. When thousands of people in an area get the same disease it is called an epidemic. If the disease spreads across the world it is called a pandemic.

You get better when your lymphocytes make enough antibodies to kill the pathogens. If you get infected again by the same disease your body already has the antibodies and so the pathogens are killed before they can make you ill. This is how your body becomes **immune** to some diseases.

6 **a** Classify the following as:
   i caused by bacteria
   ii caused by viruses.
   chicken pox, measles, tuberculosis, mumps, rubella, tetanus, dysentery, smallpox, cholera, polio, influenza

b Which of these have you had?

c Which of these have you had only once?

d Which have you had more than once?

e Which are you not immune to?

7 Bird flu started in China. It has similar symptoms to the 'flu' that you might have had. It can be fatal. There are worries that it may change and become very infectious. There are worries that it might become a pandemic.

a What are the symptoms of bird flu?

b What type of pathogen is it?

c How could it be transferred from one person to another?

d How might bird flu become a pandemic?

The fight against disease

By the end of this topic you should be able to:

- evaluate the advantages and disadvantages of being vaccinated against a disease
- explain how a better understanding of antibiotics and immunity has changed the treatment of disease
- explain how some medicines help to relieve the symptoms of infections but do not kill the pathogens causing them
- explain how antibiotics and vaccinations work.

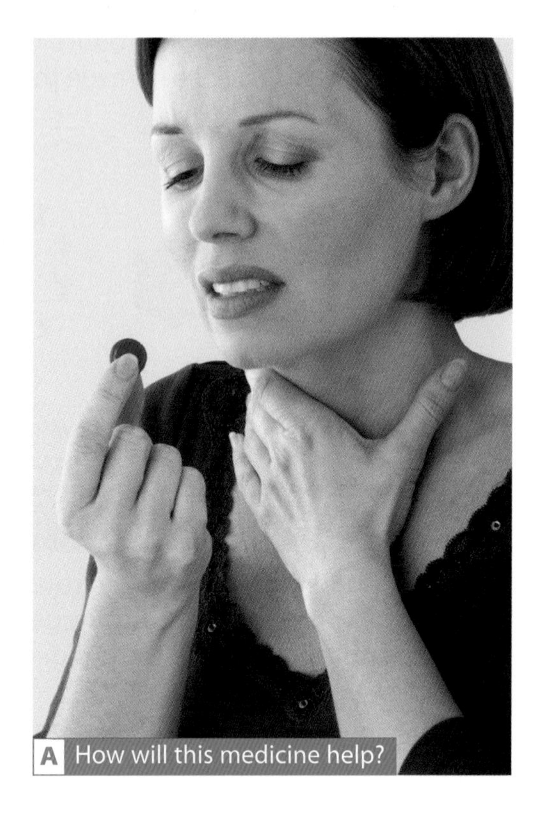

A How will this medicine help?

If you have a sore throat you might take throat lozenges to reduce the pain. The sore throat is a symptom caused by a pathogen that has infected your body. This medicine helps to relieve the symptom but it will not kill the pathogen.

1 What symptoms might you have if you catch a cold?

2 What medicines could you take to relieve cold symptoms?

If you are infected with bacteria and they multiply before you can make enough antibodies to stop them, then you become ill. A doctor might give you antibiotics.

Some **antibiotics** kill bacteria. Others stop bacteria from reproducing. Once the infection is reduced your own white blood cells can produce enough antibodies to get rid of it.

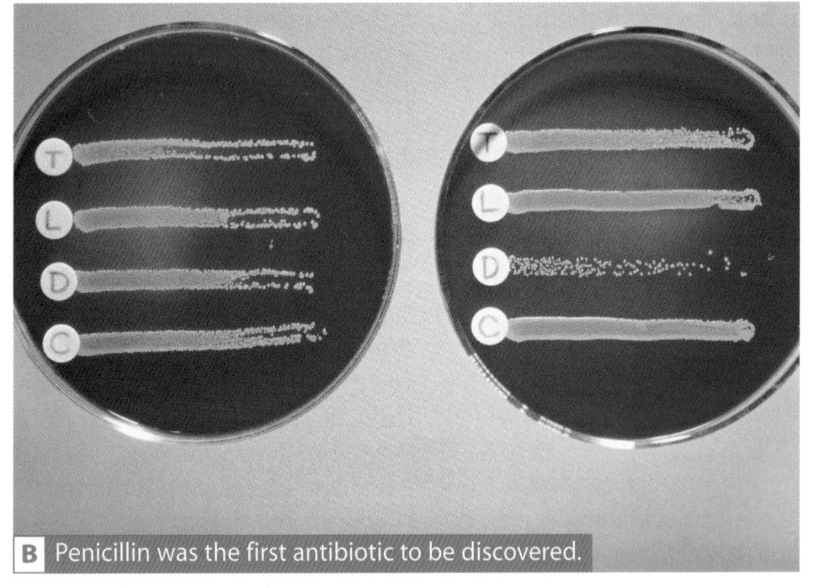

B Penicillin was the first antibiotic to be discovered.

Some organisms make chemicals to stop themselves from being attacked by pathogens. For example, one type of fungus makes a substance called penicillin. Alexander Fleming discovered that it stops bacteria from multiplying. It has been used as an antibiotic ever since.

3 If you are given antibiotics you must finish the course of treatment. Why is it important to finish the course?

4 Why is it important not to use antibiotics every time you are ill?

5 How has the development of antibiotics helped us to reduce the effects of bacterial diseases?

Antibiotics are only used to treat diseases caused by bacteria. They do not work against viruses because viruses reproduce inside cells. Viruses are difficult to target. To kill viruses you might have to destroy the cells they are in. This can do more harm than just leaving your body to deal with the virus itself.

Vaccinations are given as injections to stop you from being infected by some diseases. The fluid used is a **vaccine**. It contains a small amount of dead or inactive pathogens. Your immune system detects them and makes antibodies to destroy them. Then if you get infected by the disease, your body already has the antibodies and so you don't become ill. You are said to be **immune** to the pathogen.

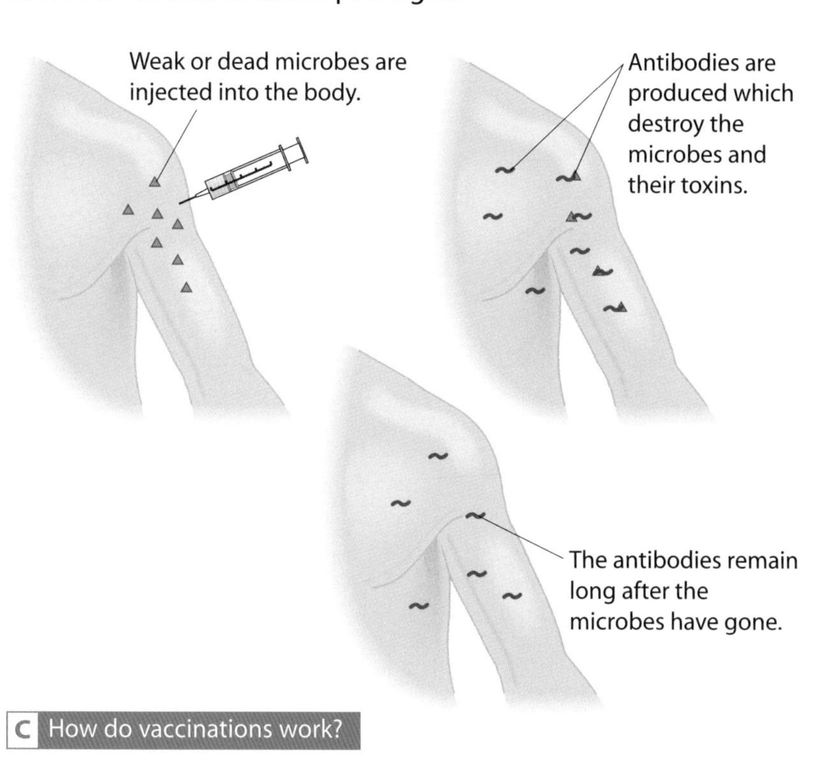

C How do vaccinations work?

MMR triple vaccine

Small children are given vaccinations against measles, mumps and rubella (German measles). It is called the **MMR triple vaccine**. As with all medical treatments, vaccinations carry a risk of possible side-effects. Some parents stopped their children from having the MMR vaccine because certain people thought there was a small risk of it causing autism, a type of brain damage.

8 Make a list of vaccinations you have had.

9 Rubella is a pathogen that can pass across the placenta. Why is it important that girls are vaccinated against it?

6 Explain why antibiotics are not used to kill viruses.

7 Why are antiviral drugs only used treat more dangerous viruses?

10 a What does MMR stand for?

b Why is it important to be vaccinated against the three diseases?

c How does the vaccine work?

d Why have some parents stopped their children having the MMR vaccine?

e How would you advise these parents? Explain your answer.

B1a.15

Drug development

By the end of this topic you should be able to:

- explain how strains of bacteria can develop resistance to antibiotics and the role of overuse in causing this
- evaluate the role of pathogen mutations in causing epidemics and pandemics
- describe how many drugs are derived from natural substances and have been known to indigenous peoples for many years
- explain how scientists test and trial drugs they develop
- describe how some drugs can be beneficial but others may harm the body (e.g. Thalidomide).

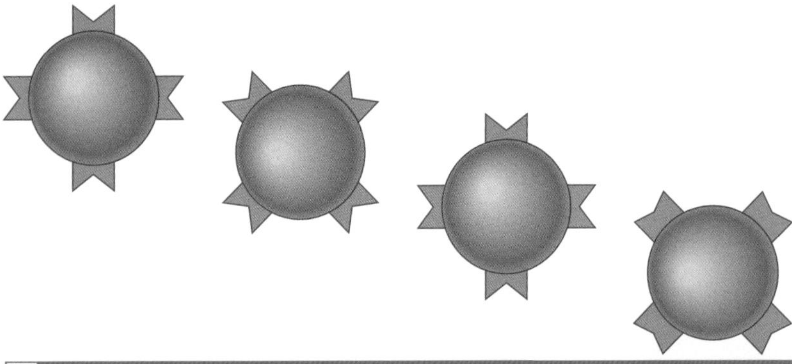

A Bacteria grow and divide every 20 minutes. Each new bacterium is exactly the same as the one it came from. Sometimes a bacterium forms that is slightly different to the others. This is called a **mutation**.

If you are infected by the bacteria in diagram A and you have the antibodies in your blood they will target and kill all the bacteria except the **mutant** form. This is an example of **natural selection**. To kill the mutant form your body has to develop a new antibody. This takes time and so you become ill. New types of the same pathogen are called **strains**.

Antibiotics which kill the normal form of the bacteria cannot kill these mutants. They are said to be **resistant** to the drug.

If we use antibiotics too often, bacteria build up a resistance to the drug more quickly. This can then lead to 'superbugs'. These are bacteria that are resistant to many antibiotics. If a pathogen cannot be controlled, it can spread through a population very quickly causing an epidemic or even a pandemic.

1 a Which of the four bacteria in diagram A is the mutation?
b Which one will not be killed by the antibody? Explain your answer.

2 One type of disinfectant claims to kill 99% of household germs. Why might this be a problem?

3 a What type of pathogens do antibiotics usually kill?
b What might happen if an antibiotic is used too often?
c How can this problem be reduced?

Clinical trials

Once a drug has been developed, it is tested in the laboratory to make sure it is not toxic. If it seems safe it is then tested on humans. The process is called **clinical trialling**.

Phase	Test	Purpose	Sample size (number of people)
Phase 1	safety and dosage tests	checks it is not toxic checks the safe amount (dosage) to give patients	20–80
Phase 2	check the drug works in humans	checks it works in humans checks for side-effects	100–300
Phase 3	random testing	checks results are reliable checks the drug works for all types of people compares with other similar drugs checks shelf life	1000–3000
Phase 4	used by the public	monitoring for any long term or rare problems that have not been picked up in phase 3	

B Clinical trials are carried out in four stages.

4 Suggest why only a few people are used in phase 1 of a clinical trial.

5 Suggest why phase 2 is carried out after phase 1, and not at the same time or before.

6 **a** What does 'random testing' mean?
b Why does the testing in phase 3 have to be random?

7 Once a drug is on sale why does it still have to be monitored?

Sometimes clinical trials go wrong. A drug called **Thalidomide** was given to pregnant women to help them sleep and to stop them feeling sick. When their babies were born some had deformed or missing limbs. The drug was withdrawn. Recently, Thalidomide has been used in the treatment of leprosy.

Many medicines are developed from plants. Aspirin was developed from the leaves and bark of the willow tree.

Malaria is a disease caused by parasites and passed to humans by mosquitoes. If a malarial mosquito 'bites' you, the parasite gets into your bloodstream. Malaria kills more than a million people a year. Some strains are now resistant to many antimalarial drugs. Recently, a new drug that comes from a plant called sweet wormwood has been developed. The plant has been used for over 1500 years by the Chinese. The parasite that causes malaria has no immunity to this drug.

There were once many indigenous peoples who knew which plants to use to treat many diseases. Many of these people have disappeared and their knowledge has been lost forever.

8 Where might the clinical trials of Thalidomide have gone wrong? Explain your answer.

9 **a** Why do antimalarial drugs not work as well as they used to?
b The World Health Organisation once promised to destroy malaria completely. Why was this not a wise thing to say?
c Why might the malaria parasite not become resistant to the new drug developed from sweet wormwood?
d Suggest why the new drug can not yet be bought.

Assessment exercises

Part A

1 Which of the following diseases is NOT caused by smoking tobacco?

a cancer
b emphysema
c bronchitis
d influenza

(*1 mark*)

2 When you get hot during exercise, which of the following does not happen?

a vasodilation
b vasoconstriction
c get hot
d sweat

(*1 mark*)

3 Which of these is an example of negative feedback?

a control of water lost through the lungs
b hypothermia
c heat stroke
d diabetes

(*1 mark*)

4 The graph shows how the number of bacteria has changed over a period of time.

a Both variables are continuous.
b Neither variable is continuous.
c Time is the only continuous variable.
d The number of bacteria is the only continuous variable.

(*1 mark*)

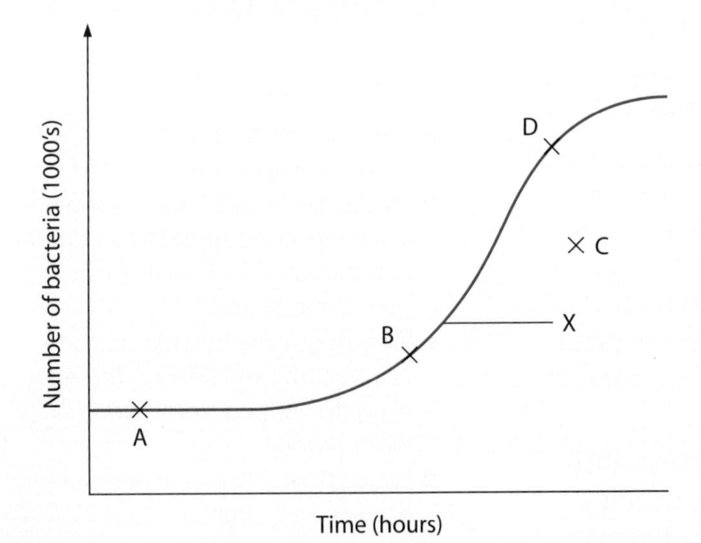

5 In the graph in question 4, which piece of data is anomalous?

a A
b B
c C
d D

(*1 mark*)

6 The diagram below shows a neurone. Match the words with their job. (*4 marks*)

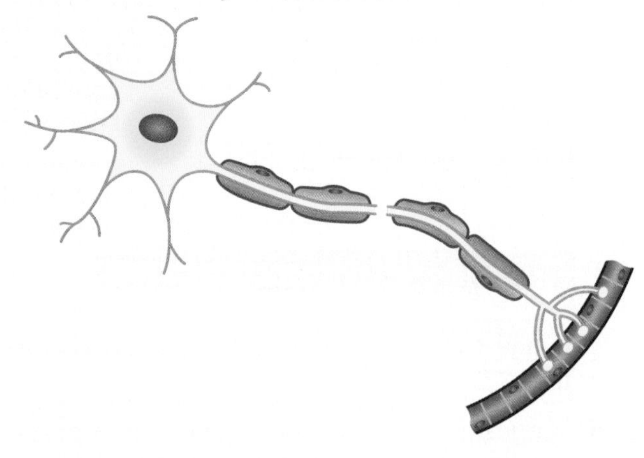

a relay neurone
b motor neurone
c sensory neurone
d receptors

i carry impulses to the spinal cord
ii carry impulses through the spinal cord and up to the brain
iii take the spinal cord to an effector
iv sensor that detects stimuli

7 Copy and complete the sequence below, showing the route an impulse takes along the nervous system, using the following words: effector, sense organ, motor neurone, sensory neurone. (*4 marks*)

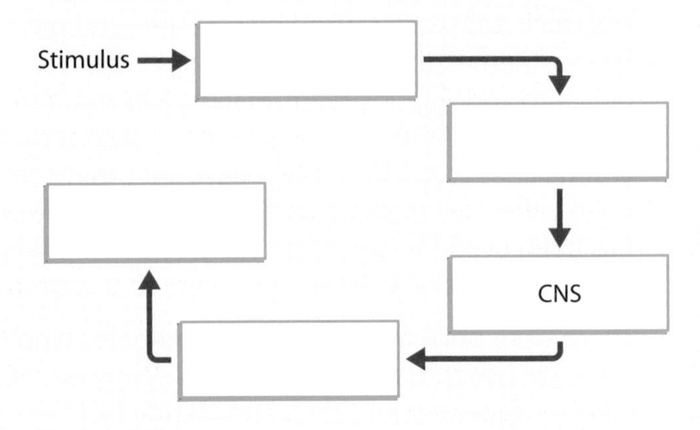

8 In the table below, match the following hormones to the glands that they are secreted from: glucagon, thyroxine, ADH, oestrogen. (4 marks)

Gland	Hormone
Pituitary	
Thyroid	
Pancreas	
Ovary	

9 The Body Mass Index (BMI) is used to tell us whether we have the correct weight for our height. Match the BMI figures **a–f**, shown below, with the spaces in sentences **i–iv**. You do not have to use all the BMI figures. (*4 marks*)

BMI

a Less than 15
b 15–19
c 20–25
d 26–30
e 30–35
f 35+

i An ideal BMI would be ____________.
ii Somebody who is obese would have a BMI of ____________.
iii Somebody who is suffering from Anorexia Nervosa might have a BMI of ____________.
iv Somebody who is a little overweight would have a BMI of ____________.

(*4 marks*)

10 Look at the data below that shows the caffeine content of some popular drinks.

a Type of drink	**b** Average caffeine content (mg) for a typical measure
Instant coffee	70
Tea	60
Cola brand 1	64
Cola brand 2	43
Cola brand 3	61
Energy drink	80
Sugar-free cola	46
Coffee (large)	500
Coffee (filter) on average	130

The variables are
a Type of drink
b Average caffeine content (mg) for a typical measure

i Which variable is categoric?
ii Which variable is continuous?
iii Which variable is discrete?
iv Which variable is dependent?

(*4 marks*)

Total (Part A) 25 marks

Part B

1 Describe the route an impulse takes from the initial stimulus to the brain. (*3 marks*)

2 a Name the addictive chemical that is found in tobacco. (*1 mark*)
b Name the gas that replaces oxygen in haemoglobin when tobacco is smoked. (*1 mark*)
c Give three ways in which a person might stop smoking. (*4 marks*)

3 Identify three things that your body does to control your internal environment when you get hot. (*3 marks*)

4 When a new drug is trialled, random tests are carried out to make sure that the drug is safe.
a What would be the dependent variable in this type of test? (*1 mark*)
b What would be the independent variable? (*1 mark*)
c What type of variable would random testing involve? (*1 mark*)
d Why is the testing random? (*1 mark*)

Total (Part B) 16 marks

Investigative Skills Assessment

Nathalie is comparing an 'off the shelf' antibacterial face-wash with a antibacterial hand wash that is used in hospitals.
She placed three discs, each loaded with one drop of the agent, onto an agar plate that had been inoculated with bacteria.

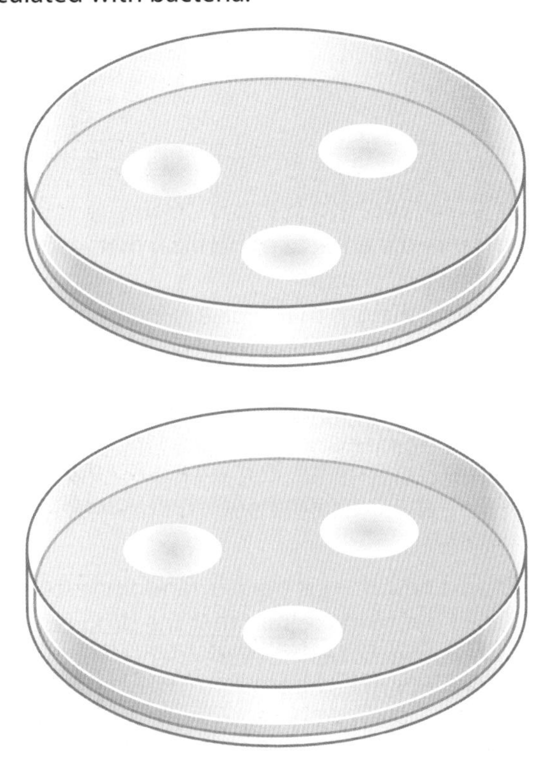

She repeated the investigation three times with each antibacterial agent.
Her results are shown in the tables below.

Face-wash	**Radius (cm)**		
	Plate 1	**Plate 2**	**Plate 3**
Disc 1	0.7	0.7	0.6
Disc 2	0.7	0.8	0.5
Disc 3	0.6	0.8	0.8
Average	0.7	0.8	

Hospital hand wash	**Radius (cm)**		
	Plate 1	**Plate 1**	**Plate 1**
Disc 1	1.6	1.5	1.2
Disc 2	1.5	1.8	1.5
Disc 3	1.2	1.7	1.3
Average	1.4	1.7	1.3

1 Complete the average clearance of Plate 3 for the face-wash. (*1 mark*)

2 Plot a bar chart showing the average clearance on each plate. (*3 marks*)

3 What is the range of data for this test? (*1 mark*)

4 What is the independent variable for this investigation? (*1 mark*)

5 Is the independent variable continuous, discrete or categoric? (*1 mark*)

6 What is the dependent variable? (*1 mark*)

7 Is the dependent variable continuous, discrete or categoric? (*1 mark*)

8 **a** According to the data which antibacterial agent is the best? (*1 mark*)
b Explain your answer. (*3 marks*)

9 Nathalie repeated the investigation using only one type of antibacterial agent. This time she altered the concentration of the antibacterial agent. What would you expect the relationship to be between the area of bacteria cleared and decreasing the concentration of the agent? (*2 marks*)

10 ✎ Discuss how Nathalie might have made sure that the investigation was a fair test. What other things might she have done to improve the investigation? (*4 marks*)

Total 19 marks

Glossary

antidiuretic hormone (ADH) A hormone that controls water levels in the blood.
antibiotic Chemical that kills bacteria and fungi, used to treat infections caused by them.
antibodies Proteins that destroy particular microbes. They are made by white blood cells.
antitoxins Chemicals that destroy toxins. They are made by white blood cells.
Bacterium (plural bacteria) A type of single-celled microorganism without a nucleus.
balanced diet A diet that has the correct amount of each type of nutrient.
BMI (Body mass index) Uses the weight and height to calculate the amount of body fat.
carbohydrates A group of foods used for energy, e.g. starch and sugars. Starchy foods that are good sources of energy and nutrients, such as bread, rice, pasta and grains are called Complex carbohydrates.
carbon monoxide A very poisonous colourless and odourless gas. It attaches onto the red blood cells and stops them carrying enough oxygen around the body.
carcinogens Cancer-forming chemicals.
central nervous system (CNS) The brain and the spinal cord.
cholesterol A fatty substance found in some foods. It can clog up arteries and lead to heart disease.
cirrhosis of the liver Damage that limits blood flow to the liver.
clinical trials Research used to find out whether a new drug is safe.
dehydrate Losing too much water from your body.
diet Everything that an animal eats and drinks.
drug A chemical that changes how our body works in some way.
drug abuse When people take too much of a drug or use it for the wrong reasons.
drug addiction When a person cannot get through a period of time without taking the drug. They are dependent on the drug.
drug dependence When the body finds it difficult to work without the drug and so the person needs to take the drug regularly.
effector A part of the body that carries out a response to a stimulus. It can be a muscle that is made to contract (tighten) or a gland that releases (secretes) a chemical (e.g. a hormone).
enzymes A substance that speeds up a chemical reaction in the body (biological catalyst).
essential fat A fat that must be eaten in the diet because it cannot be made by the body.
fats A group of foods used as a food reserve and to keep the body warm.
fibre The part of foods from plants that cannot be broken down by the body.
follicle stimulating hormone (FSH) A hormone produced by the pituitary gland that stimulates the growth of eggs in the ovaries. It also makes the ovary produce a protective covering around the egg (follicle).
gateway drug A drug is thought to lead people who use it into using harder drugs.
germs Microorganisms that cause illness or disease.
gland A cell or group of cells that produce and release substances used nearby or in another part of the body.
growth hormone A hormone that controls the size (height) of your body.
HDL cholesterol Mostly protein with only a little fat. It is often said to be 'good' cholesterol because it carries cholesterol out of the arteries and back to the liver. The higher the HDL level, the lower the risk of heart disease.
hormone A chemical 'messenger' that makes a body process happen. Hormones are secreted by glands.
hygiene Keeping things clean so that the risk of disease is reduced.
immunity When our body has enough antibodies to kill germs before they make us ill.
impulse An electrical message that is sent along the sensory neurone when a receptor senses something.
infection Once a pathogen is inside your body.
inflammation An area that becomes red and swollen as phagocytes attack the pathogens.
ion An atom or group of atoms with an electrical charge.
LDL cholesterol Mostly fat with only a little protein. It is often said to be 'bad' cholesterol because it collects in the arteries. The higher the LDL level, the higher the risk of heart disease.
lean body mass Mass that takes into account your bones, muscles and organs.
luteinising hormone (LH) A hormone that triggers ovulation and turns the follicle into a yellow body.
malnutrition When we eat too much or not enough of one or more type of nutrients.
menstrual cycle The monthly cycle from the beginning of one menstruation to the beginning of the next. The menstrual cycle includes the development and release of an egg from an ovary.
menstruation When the lining of the uterus is released and bleeding occurs from the vagina.
metabolic rate The speed at which our body converts food into energy.

metabolism Chemical reactions that produce energy for the body.
microorganism Tiny living thing that can only be seen using a microscope.
minerals Chemical elements that are essential in small amounts for living things to stay healthy. They play a very important in almost all bodily processes, including our immune, nervous and hormonal systems.
MMR triple vaccine Vaccination against measles, mumps and rubella.
motor neurones Take messages from the spinal cord to an effector.
mutation A change in a gene.
natural selection When the environment selects individuals so that only the best adapted survive and breed.
negative feedback The process which checks the levels of something and acts to reduce them if they start to rise.
nerve fibres Lots of nerve cells bundled together.
neurone Another name for a nerve cell.
nutrient A substance that a living thing needs so that it can grow healthily.
obese Someone who has too much fat in their body for their height. Someone who is obese will have a BMI greater than 30.
oestrogen Hormones that control female sexual development.
ovary Glands that secrete oestrogen and progesterone that control female sexual development; where the eggs develop.
oviduct Another name for the Fallopian or egg tube.
ovulation When an egg is released into the oviduct (egg tube) from the ovary.
pathogen Microorganism that causes illness or disease.
pituitary gland A gland attached to the brain that produces hormones. One of these hormones controls the release of eggs from the ovaries.
pneumonia An infection of the lungs.
proteins Important substances used to build muscles, repair damaged tissue and help growth.
psychological dependence Thinking that you need the drug. There is no physical dependence on the drug.
reaction time The time between you seeing something and doing something to react to it.
receptors Sensors that are found in the body that detect stimuli.
reflex action An automatic response to a stimulus, often to protect the body from harm. The impulse does not reach the brain but travels from sensory neurone, to relay neurone in the spinal cord and straight back to an effector neurone. It is called a reflex arc.
reflex arc When an impulse does not reach the brain. The impulse travels from sensory neurone, to relay neurone in the spinal cord and straight back to an effector neurone.
relay neurones Carry messages through the spinal cord, and up to and back from the brain back along the spinal cord.
resistance When an organism has protection against something.
saturated fats A type of fat that contains the maximum amount of hydrogen in its molecules. Found mainly as animal fats.
sensory neurones These carry messages from the receptor to the spinal cord. They are connected to receptors.
statin Drug used to reduce LDL cholesterol in the blood.
stimulus (plural **stimuli**) Change within the body or in its surroundings that a receptor senses, including light, temperature, sound, pressure, movement, touch and chemicals (e.g. food).
strain New type of the same germ.
synapse Tiny gap between neurones. Chemicals are released at the synapse and travel over the gap.
Thalidomide Drug that used to be given to pregnant women to help them sleep at night and to stop them feeling sick in the morning.
tolerance When a person has to use more of a drug in order to feel the same effect.
toxins Poisonous substances made by a living thing.
unsaturated fats A type of fat that does not have the maximum amount of hydrogen in its molecules.
vaccination Being given an injection of a vaccine to help the body protect itself against a disease.
vaccine Fluid containing a small amount of dead or inactive disease-causing microbes put into the body on purpose. White blood cells make the right antibodies and the person becomes immune.
virus Tiny protein-coated particle that causes disease. It enters a living cell where it reproduces, damaging the cell. It is a type of microorganism.
vitamins Chemicals from food that are essential in tiny amounts for the body to stay healthy. They play a very important part in almost all bodily processes, including our immune, nervous and hormonal systems.
white blood cells Various types of cells in the blood that help to protect the body from disease. They are carried in the bloodstream.
withdrawal symptoms The unpleasant side-effects when someone stops taking a drug they are addicted to.

Evolution and environment

A

The Eden Project was built in 2000 to encourage people to reconnect with nature and look at the world with fresh eyes. It was built in an abandoned china-clay pit and carefully designed to have minimal impact on the environment. It has large, dome-shaped greenhouses to protect those species not adapted to the British climate. It also has large outside areas planted to show plants that are important to the British countryside and to agriculture.

The Eden Project aims to show us what we should value in our living world. It also encourages us to think about how we can use the world's resources in a way that does not damage the environment so that future generations will be able to enjoy what we have now.

By the end of this unit you should:

- understand why organisms live where they do
- know why organisms of the same species are different to each other
- understand how we can produce organisms with useful characteristics
- understand why some species die out
- know how we are affecting the environment and other species by our activities.

Read these statements and sort them into the following groups: I agree, I disagree, I want to find out more:

- Adaptations to the environment help organisms to survive and breed.
- Cloning and genetic engineering could lead to advances in medicine.
- Genetically modified crops could damage the environment.
- Darwin's theory of evolution is accepted by everyone.
- We are causing a mass extinction of organisms on the Earth.
- The pollution we make is harmful to the environment.
- If countries don't develop with care for the environment, we will damage the Earth for future generations.

B1b.1

Adaptation for life

By the end of this topic you should be able to:

- explain that organisms need a supply of materials from their surroundings and other organisms so that they can survive
- suggest how organisms are adapted to the conditions in which they live.

Plants and animals are adapted to the environments that they live in.

All organisms need food, water and nutrients to grow. They get these from the environment where they live. Environments, such as rainforests and coral reefs, are very different. To grow well and produce offspring, an organism needs particular characteristics that help it get the materials it needs from its environment. We call these characteristics **adaptations**.

1 Photograph A shows plants that grow in tropical rainforests. Describe some of the environmental conditions that they need to be adapted to.

2 Photograph B shows many kinds of animal that live in warm, tropical waters. Describe at least two ways that the species here might interact with each other.

Plants in rainforests have to cope with large amounts of rain. Water is heavy and can damage leaves if it collects on them. The leaves are adapted so that water runs off them quickly, or falls through them so that they are not damaged. If you look at photograph A you will see that some leaves have shiny surfaces that help the water to run off easily. Pointed tips also help the water to fall from the leaf more quickly.

3 a Describe three adaptations of rainforest plants to their environment.
b Explain why the adaptations help these plants to survive there.

4 A rainforest leaf that is split into sections can be bigger than one that isn't split. Suggest why this is an advantage.

C This leaf is separated into sections so that rain falls through it more easily.

Many plants grow in places where there is little water. Plants like cacti have adaptations such as very deep or wide root systems to collect as much water as possible. Plants that live in dry areas also need adaptations to reduce the amount of water they lose. They lose water through tiny holes (stomata) in their leaves when they make food by photosynthesis. They need water and carbon dioxide to photosynthesise and so need the holes to let the carbon dioxide in.

D Hairy stems and no leaves reduce the amount of water a plant loses.

E A bulky body contains tissue that stores water inside the plant.

5 Describe one plant adaptation for gathering as much water as possible.

6 a List as many ways as possible that plants can reduce the amount of water they lose.
b How else can plants make sure they have water when they need it?

7 Why can't plants close the holes in their leaves (stomata) when they are photosynthesising?

8 Design a plant that would be well adapted to living in dry conditions. Draw your plant and label the adaptations that would help it to survive.

Living in cold and in dry conditions

By the end of this topic you should be able to:

- suggest how animals are adapted to survive in dry desert environments
- suggest how animals are adapted to survive in cold arctic environments.

Animals need food and water to stay alive. In dry environments there is very little water for much of the time. Some dry environments, such as deserts, can also be very hot. Animals that live in deserts need adaptations to help them survive the heat and lack of water.

A

1 Why is there very little water in a desert?

2 Herbivores are animals that only eat plants. List three problems that a herbivore would have to cope with in a desert.

Small desert animals, such as kangaroo rats, live in holes in the ground during the middle of the day. They come out to hunt for food when it is cooler. Other animals, like African moles, stay underground all the time. Being small also makes it easier for them to lose heat from their body. Small desert mammals have special adaptations in their kidneys so that they excrete urine that contains very little water.

3 Why is hunting early and late in the day an adaptation to living in the desert?

4 a Write down one advantage of living underground all the time where it is dry.

b Write down one disadvantage.

B The desert kangaroo rat does not need to drink water. It gets all the water it needs from its food.

Animals living in very cold conditions, such as in the Arctic and Antarctic, also need special adaptations. For much of the year all fresh water is frozen, so these places can also be 'deserts'.

5 Write down a good definition for the word 'desert' that includes the Arctic.

6 What conditions do Arctic and Antarctic animals need to be adapted to?

It is easier to keep warm if you have a big, bulky body because you lose heat less rapidly. Huddling together is a way of making a bulkier shape with many bodies.

Insulation also reduces heat loss from the skin. Arctic land mammals have very thick fur which insulates them from the cold winds. They may store extra fat in a layer under their skin as insulation. This can also give the animal energy when food is scarce.

7 During the Arctic summer, mammals moult a lot of their fur. Explain why they do this.

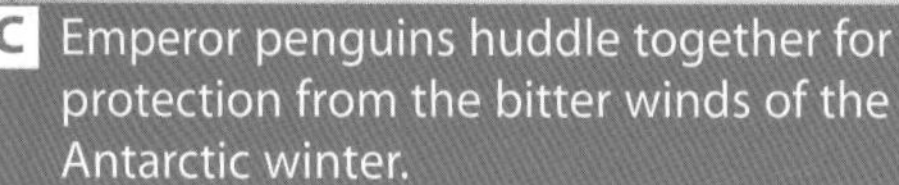

C Emperor penguins huddle together for protection from the bitter winds of the Antarctic winter.

D Musk ox in the Arctic moult some of their thick fur in the summer.

8 During the Arctic winter, a mother polar bear stays in a den with her cubs. Give as many reasons as you can why this is an adaptation to living in the Arctic.

Competing for resources

By the end of this topic you should be able to:

- explain how plants and animals compete for what they need
- describe some adaptations that organisms have to avoid being eaten.

The environment of an organism not only includes the physical conditions such as temperature and amount of water, it also includes other plants and animals. If an organism is to grow, mature and produce offspring, it must have all the resources it needs. It must also avoid being eaten by other organisms.

Plants need to get enough light and water, and enough nutrients from the soil so that they can grow well. Animals need to find enough food and water. All organisms have to compete with each other for resources.

1 Look at photograph A. Which resource are the animals competing for?

2 What will happen if they don't get enough of that resource?

A A lioness chases off a vulture from the dead antelope it was feeding on.

Animals need a mate so that they can reproduce. Males may have to compete with each other for the females, especially when they mate with a large number of females.

B The winner of this fight will be the strongest male. Only he will mate with all the females.

3 a What is the advantage for the male who wins the fight in photograph B?

b Suggest an advantage to the females who mate with the winner? (*Hint*: remember that genes from the parents get passed to the offspring.)

Genes can be transferred into a plant or animal embryo at an early stage in its development so that all the cells of the adult have a copy of the gene. For example, some crop plants like maize have been given genes for insect resistance so that the plants are not damaged by pests.

Many people are concerned that genetically modified organisms used to make our food may cause us problems when we eat them. Some people think they may also be harmful to the environment where they are growing.

5 Why are genes transferred into the cells of a plant or animal embryo at an early stage in its development?

6 Suggest an advantage of adding genes for insect resistance to maize.

C Maize is an important food crop for millions of people.

Genes can also be inserted into the body cells of an organism so that only those cells make what the gene codes for. Scientists are developing ways of treating diseases caused by faulty genes, such as cystic fibrosis, in this way. This is gene therapy.

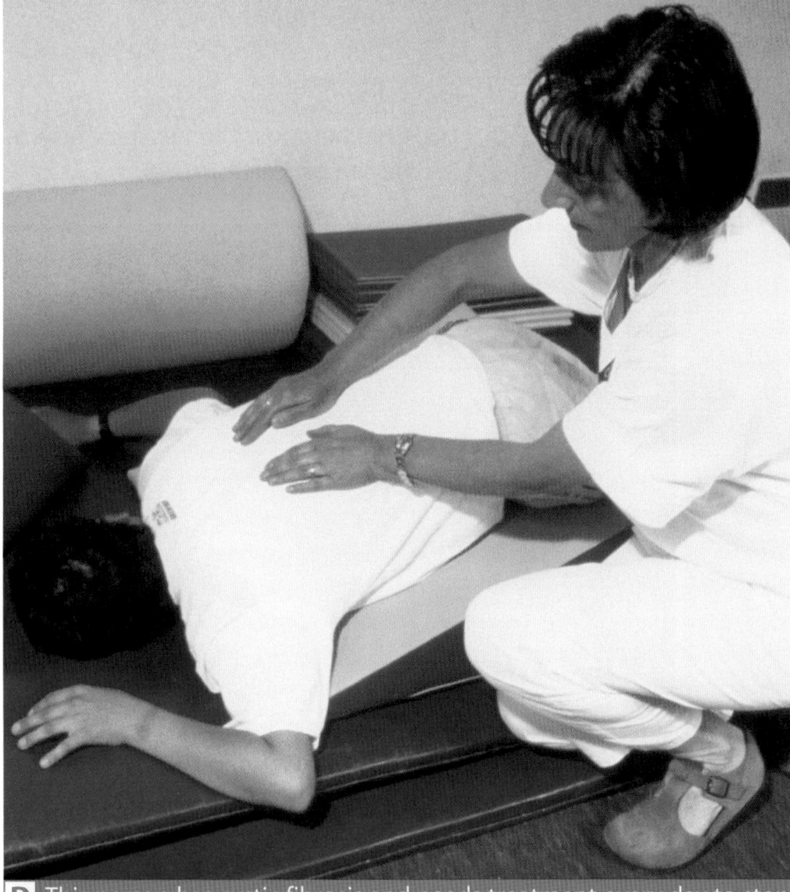

D This person has cystic fibrosis and needs treatment every day to stop him becoming very ill. In the future he might be able to breathe in the correct genes from an inhaler, but developing the treatment is proving difficult.

7 Some crop plants have been genetically modified so they contain a gene that makes them resistant to a herbicide (a chemical that kills plants). This means that the herbicide can be used to kill weeds without damaging the crop. Write a list of advantages and disadvantages of genetic modification for herbicide resistance.

8 Copy table E and complete it.

	Genetic engineering	Gene therapy
Example		
How it is done		
Which cells are affected?		

E

Life on Earth

By the end of this topic you should be able to:

- explain how fossils provide evidence of how life on Earth has changed
- suggest reasons why scientists cannot be certain about how life began on Earth
- explain that studying species helps us to understand evolutionary relationships.

The Earth formed over 4500 million years ago. The oldest rocks on the surface of the Earth are around 4000 million years old. These contain no signs of life.

The earliest signs of life are found in rocks that are just over 3500 million years old. These rocks contain shapes that look like fossils of **bacteria**. These cells may have arrived from other parts of our solar system, or they might have formed from chemical reactions that happened on Earth.

Fossils from rocks give us information about the organisms that lived on Earth in the past. Some fossils are of bones, others are just the shape of the organism or its tracks. Scientists work out what the organisms looked like by studying the fossils, and try to work out which other fossils and species on Earth today they are related to.

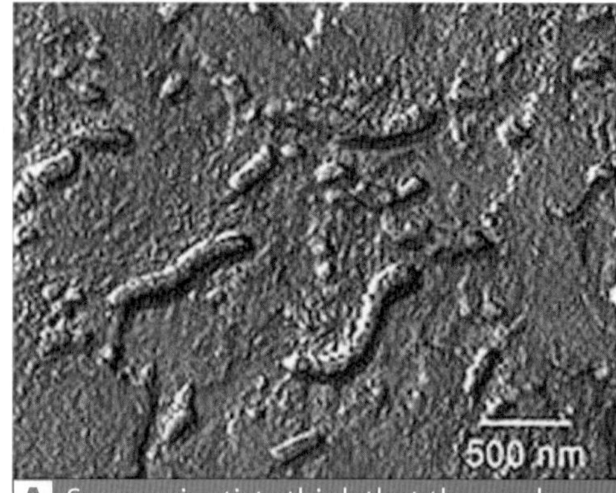

A Some scientists think that the marks on this meteorite are fossils of bacteria but other scientists disagree with this idea.

1 How old is the Earth?

2 a For how long has there been life on Earth?

b What evidence do we have to prove this?

3 Give two possible ideas for how life began on Earth.

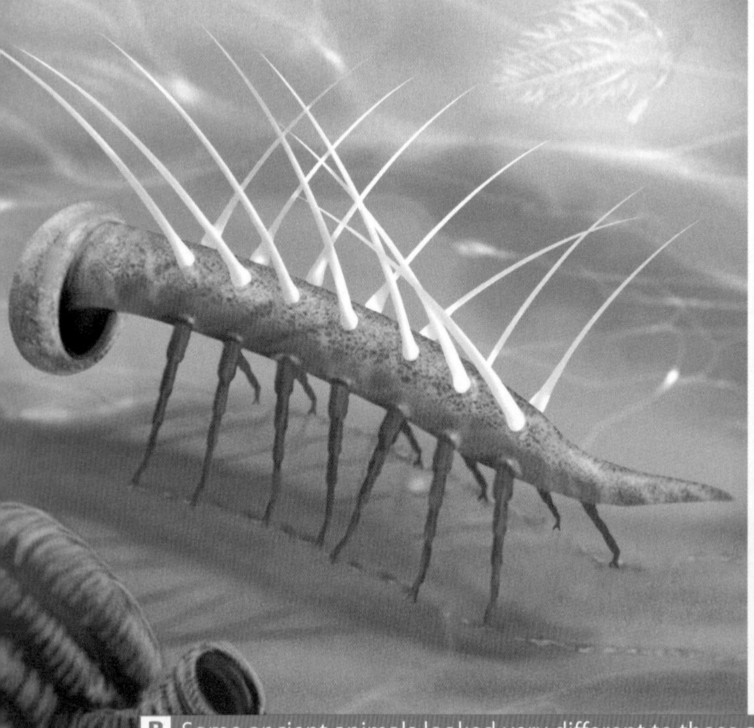

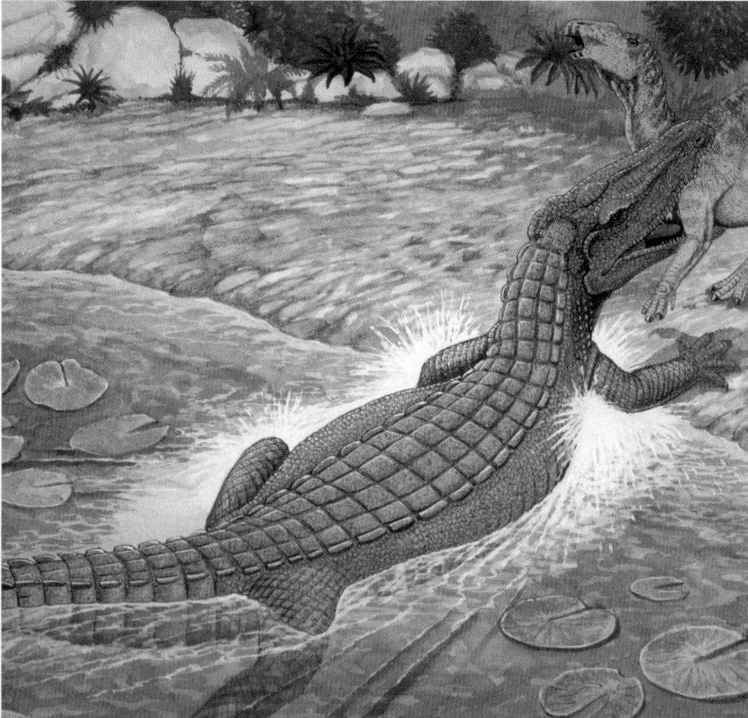

B Some ancient animals looked very different to those alive today. Others are easier to recognise. *Hallucigenia* (left) lived about 530 million years ago. *Deinosuchus* (right) lived about 70 million years ago.

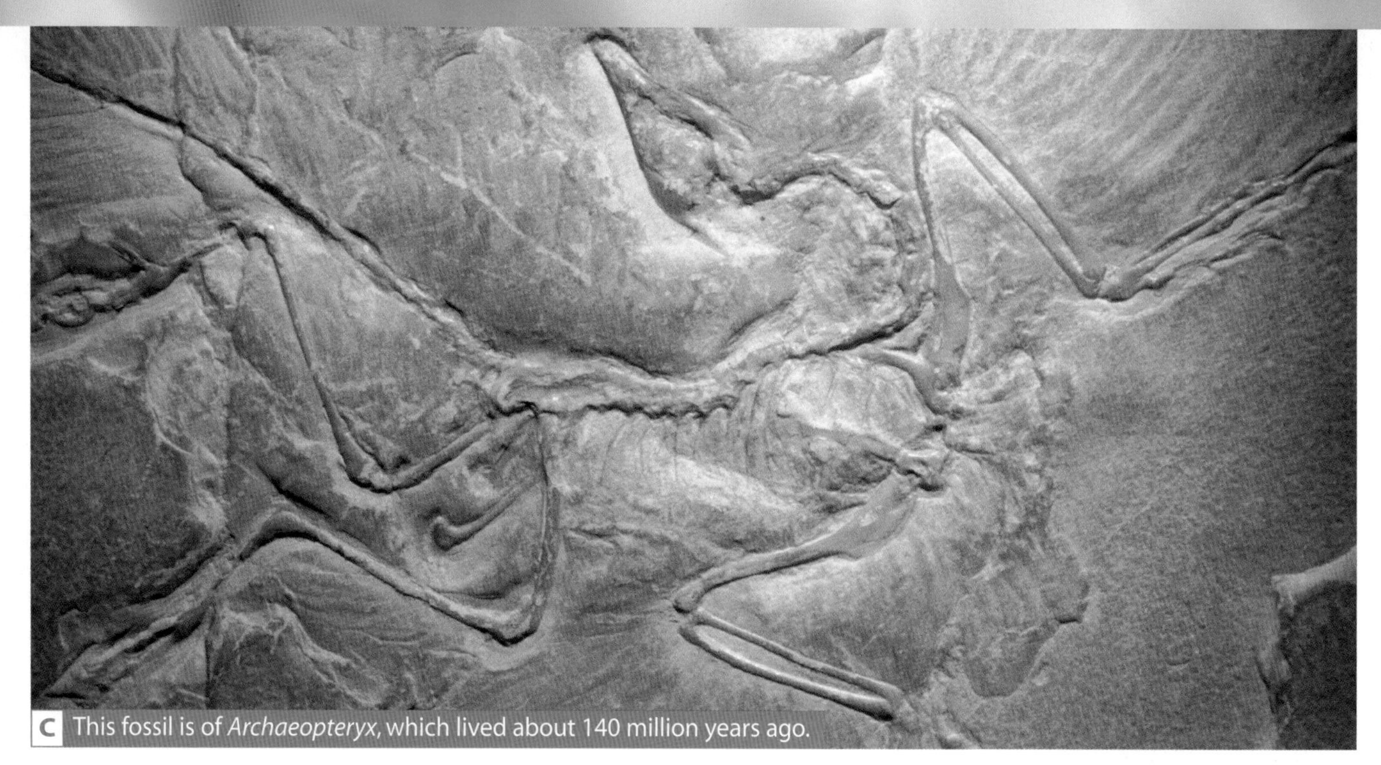
C This fossil is of *Archaeopteryx*, which lived about 140 million years ago.

Analysing rocks can tell us how old they are. If they contain fossils, this means we can work out how long ago the organisms lived.

Scientists estimate that there could be over 30 million **species** of living things on the Earth today. There are millions more species that once existed but have become **extinct**. Most people believe that all these different species **evolved** by changing gradually from the very simple cells that lived on Earth billions of years ago. We can link related fossils and living species in an **evolutionary tree** to show how later organisms may have evolved from earlier ones.

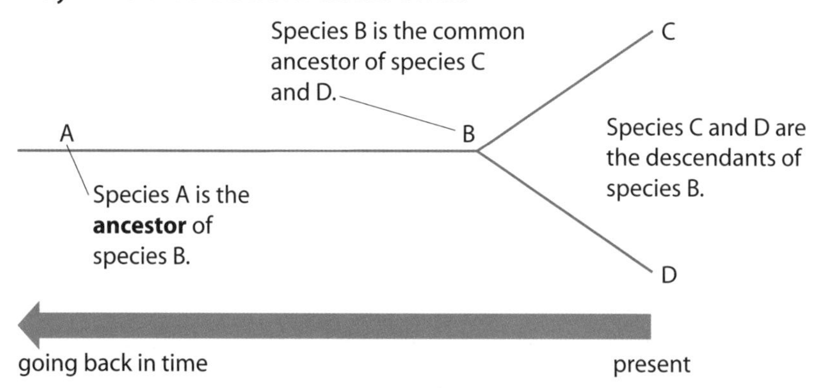

D This evolutionary tree shows that fossil organism A evolved into fossil organism B which then evolved into organisms C and D.

4 a How can scientists find out about organisms that lived a long time ago?

b How do scientists know when these organisms lived?

5 Look at photograph C. Which living animals does the fossil of *Archaeopteryx* look most like? Give reasons for your answer.

6 Why can scientists only estimate the number of species on Earth today?

7 Use diagram D to explain the meaning of the term 'common ancestor'.

8 'Life began on Earth 3500 million years ago.' How accurate do you think this statement is? Give as many reasons as possible for your answer. (*Hint*: think about the evidence we have for this.)

Natural selection

By the end of this topic you should be able to:

- identify differences between Darwin's theory of evolution and conflicting theories
- explain how natural selection can lead to evolution
- describe how mutation may result in more rapid change in a species.

Species today have different characteristics from species which lived in the past. We can see this from the fossil record. In 1809 Jean-Baptiste Lamarck suggested these changes happened when a characteristic developed (was acquired) in an organism because it was used. This acquired characteristic would then be **inherited** by its offspring.

1 According to Lamarck:
  a which is the acquired characteristic in giraffes
  b how did giraffes with very long necks evolve?

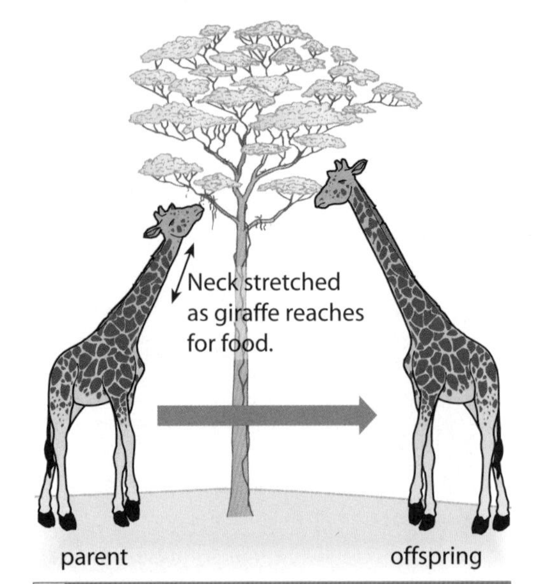

A Lamarck thought that giraffes now have long necks because their ancestors stretched to feed on tall trees and passed this characteristic on to their offspring.

In 1858 Charles Darwin proposed a different idea based on these observations:
- Most offspring produced by an organism do not survive long enough to reproduce.
- Individuals in a species show **variation**.

Darwin thought that only individuals that are best adapted to the environment survive and reproduce. So it is only these best-adapted individuals that pass their characteristics on to their offspring. Characteristics which increase the chance of survival will therefore be more common in the next generation. This idea became known as evolution by **natural selection**.

2 Suggest two things that might cause the death of an individual before it is old enough to reproduce.

3 Why are there variations between individuals?

B

4 a What do we mean by 'adapted to the environment'?
  b Explain how characteristics that increase the chance of survival become more common in the next generation.

Genes have different forms or alleles. Rapid changes in a species can happen when there is a **mutation** in an allele (a change in the genetic code). Mutations happen naturally in any cell when there is a mistake in copying the DNA. Occasionally a mutation will produce a new characteristic.

Most mutations do not affect the organism. Some cause problems so that the individual dies. A few are advantageous. For example, a mutation causes the peppered moth to be dark in colour rather than light. The moths rest on trees during the day and rely on camouflage to hide from predators.

C In sooty areas the dark form of the peppered moth is better camouflaged.

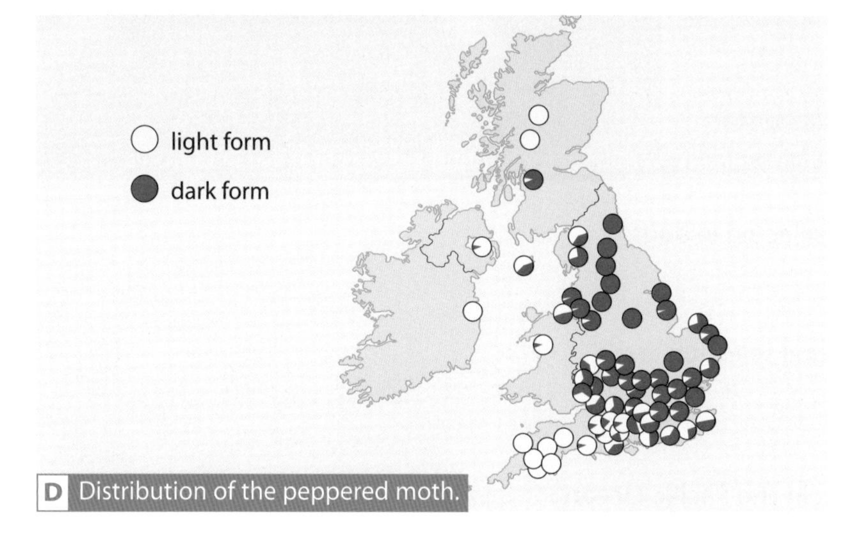

D Distribution of the peppered moth.

6 a Explain why the dark form of the peppered moth became more common when there was sooty pollution near cities.

b Predict what would happen to the proportions of dark and light moths in areas where the sooty pollution stopped.

7 How would you investigate the theory that the changing numbers of the different forms of peppered moths are due to natural selection through predation?

5 Mutations can happen in body cells and sex cells. Explain why only mutations in sex cells get passed on to offspring.

8 a Describe how Lamarck and Darwin would have answered the question 'If I do weight-training and build up my muscles, will my children inherit my big muscles?'. Include the words 'natural selection' and 'acquired characteristic' in your answer.

b Explain which answer is right and why.

Darwin's theory of evolution

By the end of this topic you should be able to:

- describe how Darwin developed his theory of evolution
- suggest reasons why Darwin's theory was only gradually accepted.

Darwin had little evidence to support his idea about evolution by natural selection. Evolution takes a long time to happen and so is difficult to see. However, Darwin knew that species change over time through **selective breeding**. Breeders do this when they select which individuals to breed from to get the characteristics they want in the offspring. Darwin used this example to persuade other scientists that natural selection could also change species.

A Wolfhounds were bred for hunting large animals. Terriers were bred for hunting animals that live in burrows.

1 Which characteristics would have been selected in dogs:
a for breeding wolfhounds
b for breeding Jack Russell terriers?

2 Explain how selective breeding and natural selection are:
a similar
b different.

Darwin had visited the Galapagos Islands in the Pacific Ocean. They are over 500 km away from South America. He suggested that a few individuals of one species of finch reached the Galapagos Islands a long while before. Over time, the birds that had beaks that were the best shape for feeding on particular kinds of food survived better than those that had beaks that were not such good shapes. They produced more offspring with the better shape of beak. Today there are 13 species of Galapagos finch, each with a different shape of beak that is adapted for feeding on different food.

B Five species of Galapagos finches. Each beak shape is adapted for eating a different food.

At first scientists found it difficult to accept Darwin's idea. This was because nobody knew about genetics and how characteristics can be passed from generation to generation. Many religious people disliked the idea because their explanation for all the different species on Earth, and all the fossils, was that God had created them.

4 Give three reasons why it was a long time before Darwin's idea was accepted.

Most people now accept Darwin's idea because over the last 150 years we have learned about genetics. There have also been many more scientific studies of how species change over time. In 1977 Peter and Rosemary Grant tested Darwin's theory. During a drought, only large seeds were left for one finch species to eat. They predicted that natural selection would mean the finches with the largest beaks would survive best and that this would mean the next generation would have slightly larger beaks. Measurements showed that the offspring had beaks that were 4% larger.

5 How does our knowledge of genetics explain how characteristics are passed on to the next generation?

6 Explain how the Grants used Darwin's theory to make their prediction.

7 Some insects burrow deep into the bark of trees. How would this affect birds that feed on these insects? Explain how you would test your prediction.

3 How does evolution by natural selection explain the evolution of the finch which can pick out insects from bark?

C Many people made fun of Darwin's ideas.

8 a Draw a flowchart that shows how a scientific idea becomes accepted. Include the words 'evidence', 'prediction' and 'theory' on your chart.

b Use examples from Darwin's idea to explain the stages in your chart.

P P A P D P P

Survival and extinction

By the end of this topic you should be able to:

- explain why a species may become extinct
- describe how humans are causing the extinction of many species.

Darwin's theory of evolution by natural selection explains why the individuals that survive and reproduce are the best adapted to the environment.

If any of the factors in the environment change, then only those individuals that are well adapted to the new conditions will survive and breed successfully. If the changes are too great or too fast, then there may be no individuals that are adapted, and they will all die. The species will become extinct.

The fossil record shows that there have been millions of species that once lived on the Earth, but are now extinct. From the rocks that the fossils were found in we can see that the conditions on Earth sometimes change a lot. This can cause a mass extinction when many species die at the same time.

The dinosaurs, and many other species became extinct as the Earth's surface cooled between about 80 and 65 million years ago. At the time there were a lot of volcanoes erupting near India.

1 What do we mean by environment? (*Hint*: remember your answer needs to include other organisms.)

A *Tyrannosaurus rex* is now extinct.

B Gases and ash from volcanoes can block out the Sun's rays and cool the Earth.

2 How could volcanoes erupting for a million years change conditions on Earth?

A species may become extinct if a competitor or predator is introduced into the area where it lives. For example, around 10 000 years ago many large herd animals in northern Europe became extinct. This links (correlates) with the time that the numbers of humans were increasing and moving into these areas.

3 a How might humans have caused the extinction of the woolly mammoth?

b Suggest another reason for the extinction of the woolly mammoth.

C This hut was made from many mammoth bones by people who lived over 10 000 years ago.

Plants and animals may also die if they become infected with a disease or a parasite (organisms that live and feed on or inside other organisms). This does not usually cause the extinction of a whole species, but it might happen when a parasite or disease is introduced to an area where it did not exist before. This is because the plants or animals have not developed resistance to the disease or parasite.

4 Extinction of species happens most easily on small islands. Suggest why.

We are changing conditions on the Earth rapidly. We are changing environments. For many species we are predators or competitors. We also move species from one region to another, and so introduce new predators and competitors.

5 Explain how we are competitors with other species.

6 Use the headings in table D to make a table of all the factors you can think of that could cause a species to become extinct.

Factors linked to non-living things	Factors linked to living things

D

The human effect

By the end of this topic you should be able to:

- describe how the rapid growth in the human population and standard of living are causing problems
- describe how these changes affect other organisms.

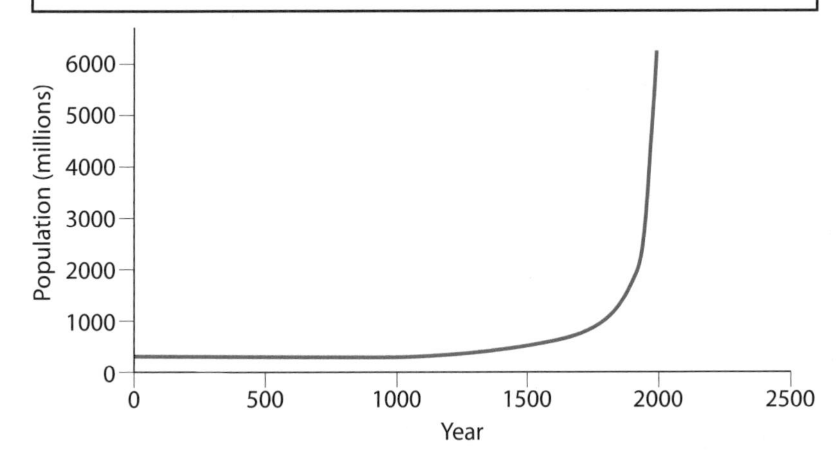

World population 1AD–2001 (UN historical estimates, 1995)

A This graph shows how the human population has increased over the past 2000 years.

Increasing population

B What do we need to live comfortably?

We are living organisms, which means we need sufficient food, water and space in which to live. In a simple life, we can use the natural resources around us, such as wood for making fire, or for making plates and cutlery. This makes little waste that does not decay quickly. If the trees that are chopped down grow again quickly, a small group of people will have little effect on the environment.

1 Look at photograph B.
  a Which resources do you think these people will use?
  b What waste will they produce?

2 a Which resources do you use?
  b What waste do you produce?

3 a Look at graph A. What does the shape of the graph mean?
  b Even if we all lived in the same way as the people in photograph B, what does the graph tell us about the effect humans are having on the world.
  c What do these changes mean for all other species?

In much of the world the way we live is changing. Our food is grown, packaged and distributed to shops in lorries for us. We usually live in large groups where water is piped to us and waste piped away. We may travel many miles every day, and we use many devices to make our lives easier. We say that our **standard of living** is rising.

4 Describe four ways that your standard of living has risen compared with the people in photograph B.

Everything that we make uses **raw materials**. Some of these raw materials, like rocks and coal, are taken from the ground in quarries. These are non-renewable resources. Other raw materials, such as wood from trees that can be regrown, are renewable resources.

5 **a** What do we mean by 'non-renewable resource'?
b What will happen to non-renewable resources eventually?

P

Houses, industry and roads all need lots of land. We also use land when we dump our waste. Quarrying destroys land, and we need large areas of land for farming. This all reduces the land that is left for other species to live in.

C Impacts of human activities.

V

6 List as many ways as you can think of that we are taking away land from the species that used to live there.

7 **a** List some of the problems that human population growth and increased standard of living are causing.
b Explain what effect this is having on other species.

P D P P

B1b.14

Dealing with waste and pollution

By the end of this topic you should be able to:

- explain how, if waste is not treated properly, it will cause pollution
- describe how water can be polluted
- describe how some animals can be used as indicators of water pollution.

A Where will this rubbish end up?

The UK produces about 35 million tonnes of household refuse every year. Much of this is dumped in **landfill** tips, usually on old industrial land. The rubbish is covered over with soil and left to settle.

As the waste decays in landfill tips, **methane** gas is produced. This needs to be piped away to prevent fires or explosions from happening. The gas can be used as a fuel to heat homes and offices. The waste also produces **toxic** liquids that drain into the ground. The tip usually has a lining to help stop this happening.

Incinerating (burning) refuse releases toxic chemicals. These chemicals have to be cleaned out of the gases before they are released into the air. The heat from the burning can be used to make electricity.

Some waste can be recycled and used to make new products so we don't use more raw materials. However, it costs money to collect and sort all the waste before it can be recycled.

1 How do we dispose of household refuse?

2 Why do you think old industrial land is used first for landfill tips?

3 For each of the following give at least one advantage and one disadvantage:
a disposing of refuse in landfill tips
b incineration of refuse
c recycling refuse.

Waste from industry may be even more toxic than domestic waste. Poisonous metals like mercury and lead need to be removed from the waste before disposal so that they do not pollute the environment.

4 How can industrial waste pollute the environment?

Farming can use large amounts of chemicals, including **pesticides** and **herbicides**. Some of these chemicals end up in the soil. Some break down into other chemicals quickly. Others may last and pollute the soil for many years and continue to harm plants and animals.

C In the 1960s and 1970s DDT was used as a pesticide. It is long lasting and caused numbers of many bird species to fall rapidly, including the peregrine falcon.

5 a What is the benefit to the farmer of using a long-lasting pesticide?
b What is the risk to the environment of using a long-lasting pesticide?

The **fertilisers** used in farming can be washed into waterways from the land. They can pollute streams and rivers killing many organisms. Only a few invertebrates can survive in these conditions. These invertebrates are **indicators** of water pollution.

6 a Why do farmers use fertilisers on crops?
b What could farmers do to avoid polluting water?

We also produce **sewage** from our homes and industry. If the sewage is not treated properly, it can also pollute water.

B This waste is poisonous to plants and could poison animals too.

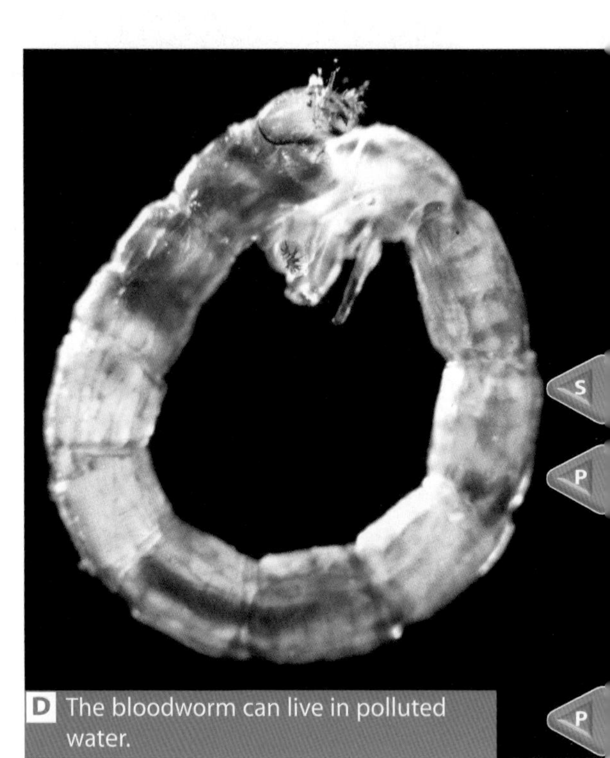

D The bloodworm can live in polluted water.

7 a List all the ways that we pollute land and water.
b How can each kind of pollution be prevented?

B1b.15

Polluting the air

By the end of this topic you should be able to:

- describe how we cause air pollution and the damage it can do
- explain how lichens can be used as air-pollution indicators
- describe how cattle and rice fields release methane gas.

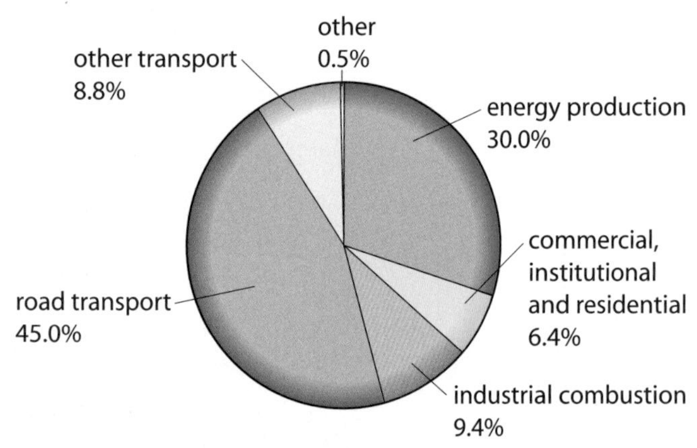

(Source: National Atmospheric Emissions Inventory)

A The total emissions of nitrogen oxides is 1.58 million tonnes.

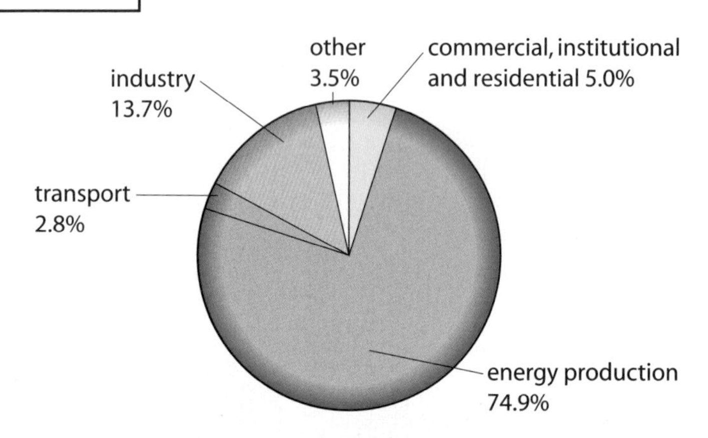

(Source: National Atmospheric Emissions Inventory)

B The total emissions of sulfur dioxide is 1.002 million tonnes.

We release smoke and gases into the air from many activities. Burning fossil fuels, such as coal, releases sulfur dioxide and nitrogen oxides. We burn fossil fuels in power stations to make electricity, in our homes to give heat, in vehicle engines and in many industrial processes.

Sulfur dioxide and **nitrogen oxides** dissolve easily in water to form acids. These acids easily damage living tissue. Gases released into the air dissolve in the water droplets in clouds and fall as **acid rain**. Rain is normally slightly acidic with a pH of around 6, but acid rain can have a pH as low as 2.

C These trees were killed by acid rain.

1 Write down the names of two polluting gases.

2 How are these gases released into the atmosphere?

3 Look at diagrams A and B.
  a What releases the most nitrogen oxides into our air?
  b What releases the most sulfur dioxide?

Acid rain damages plants and animals when it falls on them. It also makes lakes and rivers more acidic. The polluting gases can move many miles in the air before they fall as acid rain.

Lichens are plant-like organisms and there are many species. Different species tolerate different levels of acidity. So we can use them as indicators of air pollution.

4 Describe two ways that acid rain can cause damage.

5 Explain how air pollution in the UK can cause damage in nearby countries.

D The lichen on the left dies quickly if there is air pollution. The lichen on the right can tolerate sulfur pollution.

Sulfur dioxide emissions in the UK dropped by 75% from 1987 to 2003. This was because the Government introduced new laws. These laws said that industries must clean sulfur dioxide and nitrogen oxides out of waste gases so that they are not released into the atmosphere. Limits on the amounts of sulfur in fuel oil were also introduced.

Nitrogen dioxide emissions from industry have also decreased, but emissions from traffic have increased. So, levels of nitrogen oxides in the air have hardly changed.

6 How has changing our laws improved the quality of our air?

7 Look back at diagrams A and B. What else could we do to improve the quality of our air?

We release other polluting gases into the air. Carbon dioxide is a waste gas from many processes including respiration and burning fossil fuels. Methane is produced when plant material is broken down with little oxygen present. This happens inside the gut of animals, such as cows, or in soils that are flooded, such as rice paddy fields.

8 Describe how the increase in the human population is leading to the release of more carbon dioxide and methane into the air.

9 Draw and complete a table with these headings: 'Polluting gas', 'What releases it into air?', 'Is it increasing or decreasing?', 'How can we reduce it?'

A

S

P

P

D

P

P

B1b.16

Cutting down the world's forests

By the end of this topic you should be able to:

- explain why there is large-scale deforestation in tropical areas
- describe how large-scale deforestation affects carbon dioxide in the air and biodiversity.

A These photos were taken of the same part of the Amazon rainforest in South America in 1990 and 2000. The pale-coloured areas show where trees have been cut down.

Every year in the Amazon basin, an area of forest the size of England and Wales is chopped down. Some of this **deforestation** is to clear the land to make homes for people who used to live in crowded cities. They need the space to live and grow crops that they can eat and sell. Large areas of rainforest in Malaysia are also being cut down. Malaysia depends on exporting tropical timber to other countries for income.

The forests are home to many native people. There are also many species of plant and animal that are found nowhere else in the world. They depend on the forest trees for survival. These species will be lost if the trees are cleared and **biodiversity** will be reduced. Some of these species might be useful to us in the future for making medicines.

1 Name two places in the world where there are tropical rainforests.

2 **a** What is happening to these rainforests?
b Why is this happening?

3 **a** What are the benefits of deforestation to the countries where it is happening?
b What problems does deforestation cause?

Trees are plants so they make their food by photosynthesis. They combine carbon dioxide and water to make sugars. Trees are sometimes described as **carbon dioxide stores** because the gas is locked away in other chemicals for many years. This is because most of these sugars become plant tissues and remain in the tree until they are broken down during respiration or when the tree dies. Microorganisms break down the dead tree and respire, releasing the carbon dioxide.

4 What is meant by a carbon dioxide store?

5 a Write down three ways that carbon dioxide is released from a tree.

b (i) Sketch graph axes with an *x*-axis of time, and a *y*-axis of amount of carbon dioxide in the air.

(ii) Draw a line to show what happens to carbon dioxide in the air as a tree grows and is then burnt.

(iii) Label your sketch to show what is happening to the tree.

B Burning also releases carbon dioxide from trees.

Impact of deforestation

Globally, many people are concerned that large-scale deforestation will mean that there is more carbon dioxide in the air. Others believe that the natural processes will remove more carbon dioxide and that there will be little change in the atmosphere.

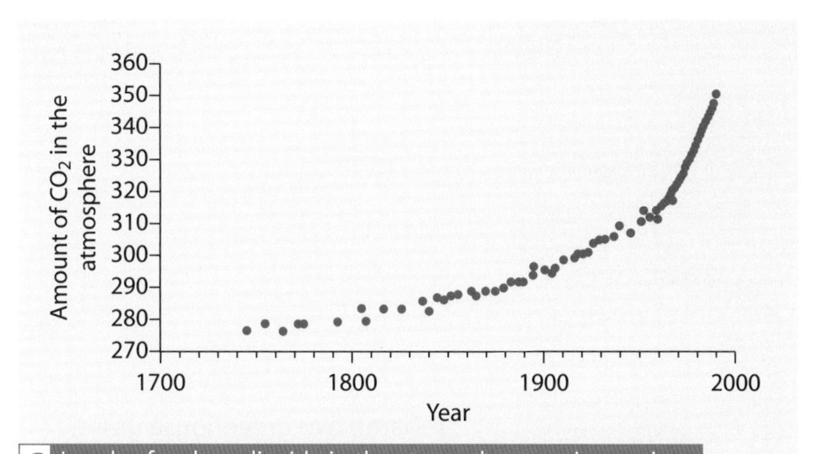

C Levels of carbon dioxide in the atmosphere are increasing.

6 Look at graph C and picture D. Explain why not everyone believes carbon dioxide levels will continue to rise in the future.

D Some other views about increasing carbon dioxide levels.

7 Write a letter to a newspaper describing what effects large-scale deforestation is having on biodiversity and the Earth's atmosphere.

B1b.17

Global warming

By the end of this topic you should be able to:

- explain how carbon dioxide and methane are involved in the greenhouse effect
- explain how increased carbon dioxide and methane may lead to global warming
- describe how global warming may result in changes to the Earth's climate and sea levels.

A Without the greenhouse effect, the Earth would be too cold for life.

The Earth receives heat from the Sun. Much of the heat that reaches us is reflected back from the Earth's surface. Gases in the atmosphere absorb most of this heat. If they didn't, it would be lost back into space. The gases then radiate the heat energy back to the Earth's surface again. These gases include carbon dioxide and methane. They are called **greenhouse gases**. They keep the surface of the Earth warmer than it would be without them. This is known as the **greenhouse effect**.

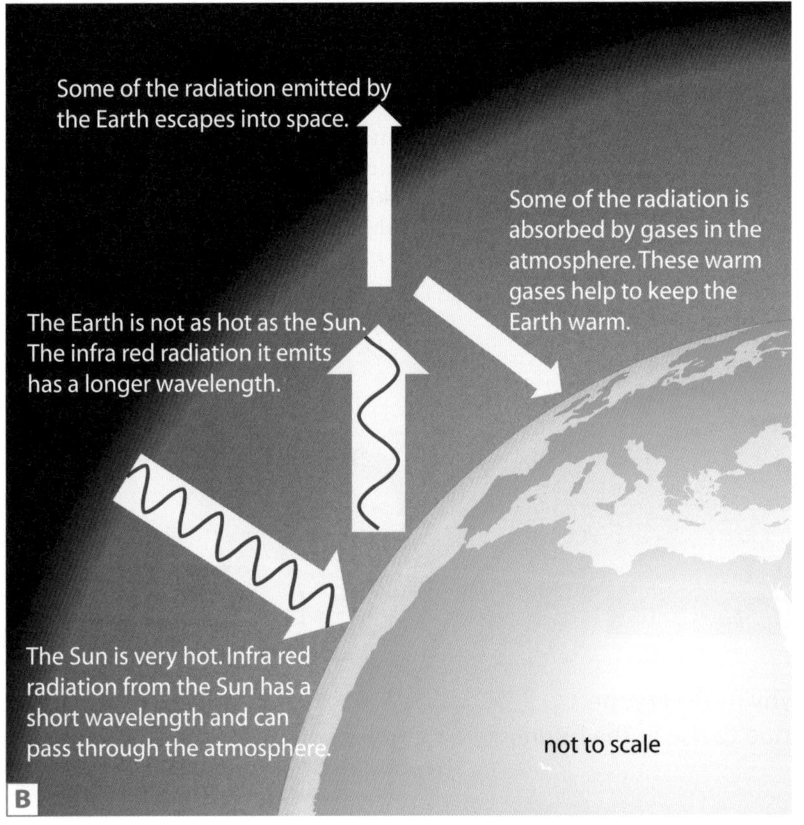

B

Increasing amounts of carbon dioxide and methane in the atmosphere increase the amount of heat reflected back to the Earth's surface. This makes the Earth even warmer. This increased greenhouse effect is known as **global warming**. Global warming is happening, and there is evidence that at least some of it is due to human activities.

1 Name two greenhouse gases.

2 Explain how they cause the greenhouse effect.

3 Why is the greenhouse effect important for life on Earth?

4 a Describe two human activities that increase the amount of carbon dioxide released into the atmosphere.

b Describe one human activity that increases the amount of methane in the atmosphere.

Computer modelling of the climate

Scientists use computer models to predict what will happen to the Earth's climate if global warming continues. There are so many different factors to take into account that there is still a lot of disagreement about what will happen. However, some people think that a rise of only a few degrees could change the way that weather patterns form. This could cause higher rainfall and floods in some places, and droughts in others. It would also change where different plants and animals could live.

Increasing temperatures are causing glaciers and polar ice caps to melt and release more water into the oceans. Some scientists predict that this could result in a sea-level rise of up to 80 metres. Millions of people live on land that is less than 20 metres above sea level. Water melting from the ice on Greenland could also stop the Gulf Stream. The Gulf Stream keeps the UK warm compared with other places which are the same distance north, such as Russia and Canada where it is much colder in winter.

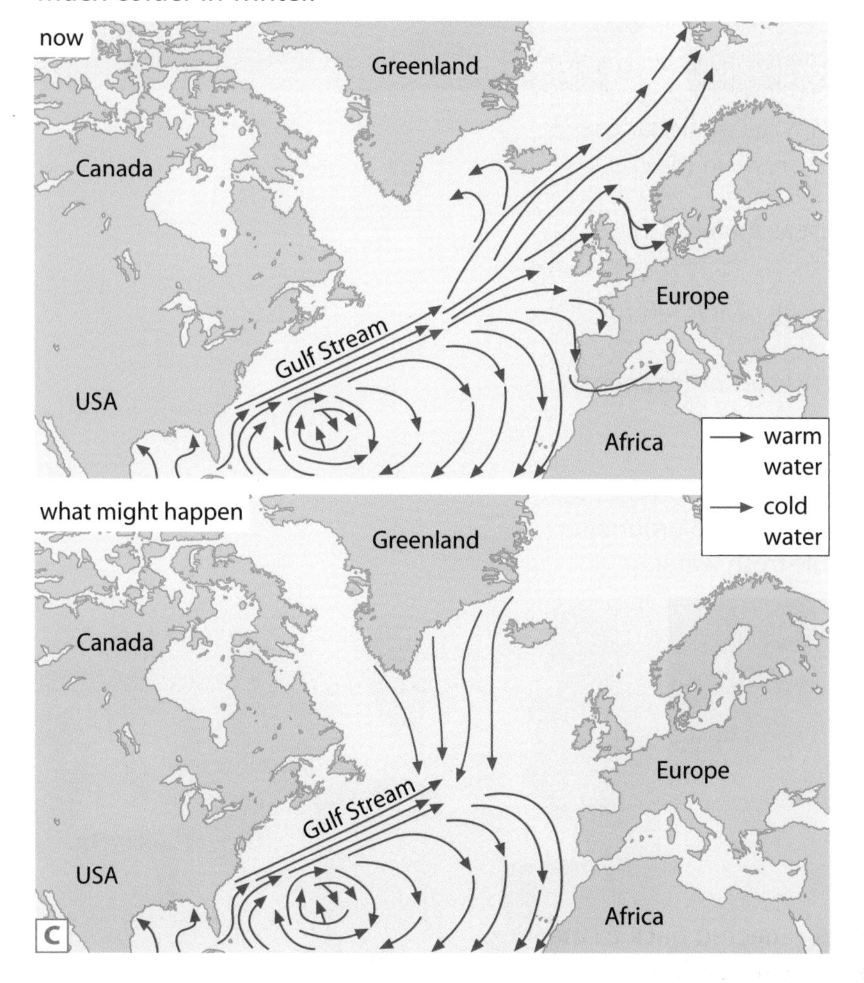

6 Explain how a small change in conditions in one place can cause a major change in conditions somewhere else.

5 What effects could changing weather patterns have on where people live? Give reasons for your answers.

7 Draw two diagrams of the Earth: one with an atmosphere and one without. Add notes to explain why one would be a colder planet than the other.

A sustainable world

By the end of this topic you should be able to:

- explain what we mean by sustainable development
- describe ways in which we can develop a sustainable world.

A We use thousands of acres to grow crops such as wheat.

B Harvesting the sea.

Each year we harvest about a third of all plant material that grows on Earth. We also take about a third of what grows in the oceans.

1 We harvest plants for many purposes. List as many of these purposes as you can.

2 a How much plant material that grows each year is left for other animals?

b Is all the plant material that is left suitable for other animals to eat? Give a reason for your answer.

There is a lot of water on Earth but not all of this is fresh water that we can drink. Some fresh water is also too deep in the ground for us to reach. We use about 60% of all available fresh water.

C We need water for more than just drinking as these two examples illustrate.

3 List all the things that you do each day that use water.

4 Two-thirds of the Earth is covered in sea water. Why is this water not as useful as fresh water?

The human population is increasing. There are about 6 billion people on Earth. By 2050 there may be 9 billion. We already have a great impact on the environment, so will things just get worse? Can we do something about this?

Many people think we can make things better by creating a **sustainable** world. This is one where we take what we need without damaging the environment for the future, either for ourselves or other species. We would need to change a lot of what we do to create a sustainable world.

Problem	What we can do
Space for everything	Only build on land that has been used before. Leave nature parks and reserves to protect wildlife.
Energy use	Use energy sources that do not pollute the environment. Become more energy efficient.
Water for everyone	Use water wisely, don't waste it. Develop ways of making sea water into fresh water.
Feed the world	Develop crops and farm animals that produce more food without taking up more land.
Pollution	Make sure we clear up all pollution.
Resources	Increase the amount we reuse and recycle resources.

D

5 Think about what you do every day and what impact your actions have on the environment. Remember to think about the things you use and how their manufacture might affect the environment.

a Make a list of five things that you do that have an impact on the environment.

b For each thing on your list, write down what the impact on the environment is and what you could do to make the action more sustainable.

c Say how easy or difficult it would be to carry out each of your sustainable actions.

6 Copy and complete table D, but with an extra column, as in table E. Look back through the last few topics and add more ideas of what we can do to be more sustainable.

Problem	What we can do	Example

E

P V V P S D P P

Assessment exercises

Part A

1 The transfer of DNA from one species to another is called:
- **a** natural selection
- **b** genetic modification
- **c** inheritance of acquired characteristics
- **d** mutation.

(1 mark)

2 Which feature of cattle could NOT be altered by selective breeding?
- **a** Sex of the offspring.
- **b** Amount of meat.
- **c** Production of milk.
- **d** Ability to produce twins.

(1 mark)

3 The table shows the approximate world human population at various times since 1650.

Year	Population (millions)
1650	500
1850	1000
1900	1500
1925	2000
1950	2500
1960	3000
1975	4000
2000	6000

If you plotted a graph of the world population against time, which of these graphs would it most closely resemble?

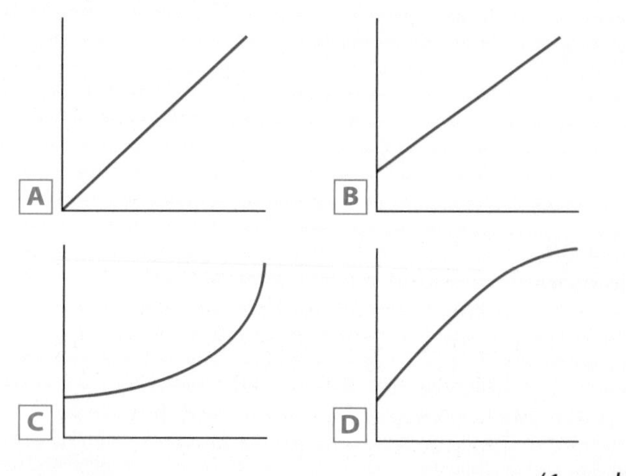

(1 mark)

4 A student wanted to find out if pollution by copper affected the growth of plants. She placed moist cotton wool in six dishes and added 10 cress seeds to each. She added five drops of dilute copper sulfate solution to the first dish, four drops to the second, three drops to the third, two drops to the fourth, one drop to the fifth and a drop of distilled water to the last dish. She placed all the dishes under the same bright light for five days, at a constant temperature of 20°C. After this time, she measured the length of the cress shoots.
Which is the independent variable in the experiment?
- **a** The length of the cress shoots.
- **b** The number of cress seeds in each dish.
- **c** The concentration of copper sulfate.
- **d** The temperature.

(1 mark)

5 In the experiment described in Question 4, which of the following would NOT have improved the reliability of the student's results?
- **a** Add measured volumes of copper sulfate solution
- **b** Increase the number of dishes of seeds
- **c** Measure the mass of the seedlings
- **d** Grow the seedlings at a different temperature

(1 mark)

6 This question is about inheritance. Match words **a**, **b**, **c** and **d** with the spaces in sentences **i–iv**.
- **a** Allele
- **b** Chromosome
- **c** Clone
- **d** Gamete

- **i** ____________: an individual with the same genes as its parent.
- **ii** ____________: a structure in a cell containing many genes.
- **iii** ____________: a form of a gene.
- **iv** ____________: a cell carrying only one parent's genes.

(4 marks)

7 This question is about evolution. Match words **a**, **b**, **c** and **d** with the spaces in sentences **i–iv**.
- **a** selection
- **b** evolution
- **c** mutation
- **d** variation

i The process that introduces new characteristics is called ___________.
ii The range of characteristics shown by a species is called ___________.
iii The process that allows the best adapted individuals to survive is called ___________.
iv The way species change over time is called ___________.

(4 marks)

8 Human activities produce a great number of pollutant chemicals. Match words **a**, **b**, **c** and **d** with the spaces in sentences **i–iv**.

a Carbon dioxide
b Methane
c Nitrogen dioxide
d Sulfur dioxide

i ___________ is a contributor to acid rain and is largely produced by burning coal.
ii ___________ is produced by flooded fields and landfill sites
iii ___________ is a contributor to acid rain that is released in engine exhaust gases.
iv Releasing large amounts of ___________ from combustion could cause global warming.

(4 marks)

9 The words below are all used in modern cloning techniques. Match words **a**, **b**, **c** and **d** the spaces in sentences **i–iv**.

a tissue culture
b embryo transplantation
c cell fusion
d adult cell cloning

i You grow small groups of plant cells to make a callus in ___________.
ii Replacement of a cell nucleus with another nucleus is called ___________.
iii If a replacement nucleus is from a skin cell, this would be an example of ___________.
iv Splitting the cells of an embryo and placing them in the wombs of different host mothers is called ___________.

(4 marks)

10 An investigation was carried out to find the effect of nitrate fertiliser on the growth of plants in an area of grassland. One area was treated by spraying with a solution containing nitrate and a second area was treated by spraying with the same volume of distilled water. Spraying was repeated at weekly intervals, for six weeks. At the end of the treatment a sampling square (quadrat) was used to sample from the treated and untreated areas. The mean height of plants in each area was measured and the number of species in each area was compared.
Match words **a**, **b**, **c** and **d** the spaces in sentences **i–iv**.

a controlled
b dependent
c discrete
d independent

i The height of the plants is an example of a ___________ variable.
ii Addition of nitrate or distilled water is an example of an ___________ variable.
iii The volume of nitrate solution or water is an example of a ___________ variable.
iv The plant species fit into categories, so they are an example of a ___________ variable.

(4 marks)

Total (Part A) 25 marks

Part B

1 The drawing shows part of a tapeworm. Tapeworms are parasites. They live inside the intestine of many species of animals, including humans. They feed by absorbing products of digestion from their host's intestine. Segments of the tapeworm pass out of the host's body with the host's waste products.

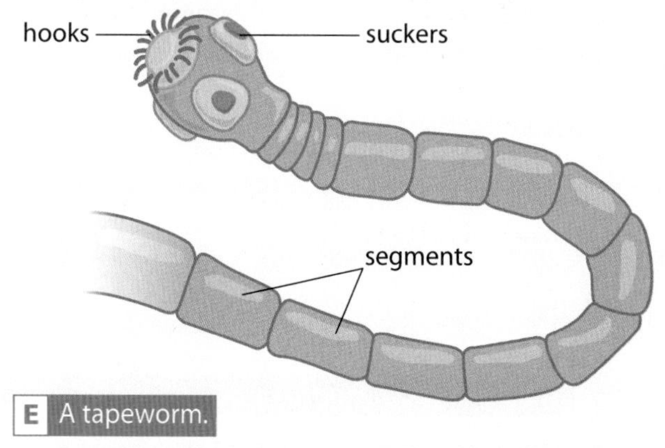

E A tapeworm.

The tapeworm shows several adaptations to living as an internal parasite.

a What do we mean by an adaptation? *(1 mark)*
b Explain how each of the following is an adaptation to life as an internal parasite.
  i Hooks and suckers on the head. *(1 mark)*
  ii Each segment contains both male and female sex organs. *(1 mark)*
  iii The tapeworm has no intestine of its own. *(1 mark)*

2 Scientists can shoot genes into a plant using a gene gun! This is a machine that fires tiny pellets made of gold. The pellets are coated with DNA containing a gene from another species, which the scientists want to transfer. It is easiest to shoot these pellets into young delicate plant tissue. Some of the cells will take up the DNA, and can be cultured and grown into adult plants to give lots of identical plants.

- **a** What is the name given to a process that transfers genes from one organism to another of a different species? (*1 mark*)
- **b** Suggest how the cells could be cultured to grow into plants. (*1 mark*)

One gene that can be transferred to crop plants gives the plants resistance to herbicides. The herbicides can then be used to kill weeds without killing the crop plant.

- **c** Some people object to this, because they say that the gene might 'escape' into other plants, including weeds, making them resistant to herbicides too. Can this be described as an **ethical** issue? Explain your answer. (*2 mark*)
- **d** Give two other characteristics, apart from herbicide resistance, that could be transferred into crop plants. (*2 mark*)

3 The diagrams show the front limb bones of three animals: horse, human and whale. They all have a bone structure called the **pentadactyl** (five fingered) limb. They have all evolved from a **common ancestor**.

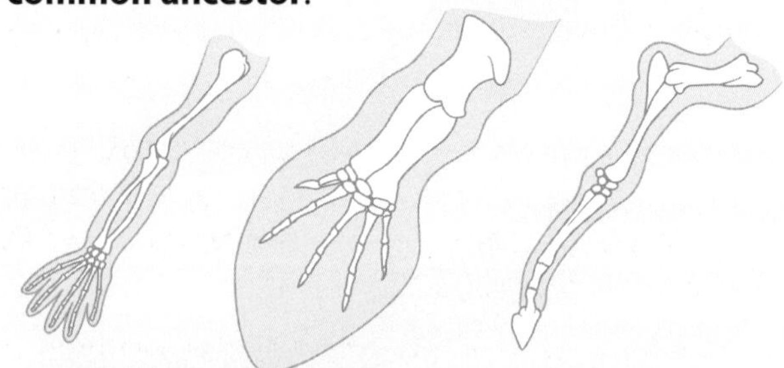

F The pentadactyl limb in three different animals.

- **a** What does a **common ancestor** mean? (*1 mark*)
- **b** Suggest which limb bones belong to which animal. (*1 mark*)
- **c** How are the limb bones of the whale adapted to its habitat? (*1 mark*)
- **d** The natural habitat of the horse is on grassy plains. It needs to run fast to escape from predators. Suggest how the pentadactyl limb has evolved to become adapted in the horse to suit this lifestyle? (*1 mark*)

4 A scientist made the following statement:

> **Deforestation** damages the environment, destroying endemic **species** and habitats, and reducing **biodiversity**. Trees may live for hundreds of years, so they act as **carbon dioxide stores**. Therefore, burning trees contributes to the **greenhouse effect**.

- **a** Explain all the terms in the statement that are written in **bold**. (*5 marks*)
- **b** Is it true to say that 'burning trees contributes to the greenhouse effect'? Explain your answer. (*2 marks*)

5 Swimming mayfly nymphs are the larvae of mayflies. They only live in streams with clean water, where there is a high concentration of oxygen. A student investigated the distribution of swimming mayfly nymphs in his local stream. He sampled from different areas of the stream and recorded the number of nymphs in areas of fast-flowing and slow-flowing water, and also measured the flow rate (water speed). He took care to sample at the same time of day, using the same technique for each sample taken. Here are his results.

Sample number	Flow rate (ms^{-1})	Number of nymphs
1	0.2	3
2	0.6	6
3	0.9	8
4	0.3	3
5	1.9	9
6	1.5	11
7	0.2	5
8	1.0	6
9	1.4	8
10	1.7	12

- **a** Plot a graph of the number of nymphs against the flow rate of the stream. (*2 marks*)
- **b** Draw a 'best fit' line through the points. (*1 mark*)
- **c** What was the student trying to find out? (What was his hypothesis?) (*1 mark*)
- **d** Do the results support his hypothesis? (Explain your answer.) (*2 marks*)
- **e** The student decided that the flow rate was affecting the oxygen content of the water. Is he justified in concluding this from the data that he has collected here? If not, what does he need to do next? (*2 marks*)

Total (Part B) 29 marks

Investigative Skills Assessment

These questions are about Pooja's investigation into the germination of cress seeds. You should use her results below, as well as your own understanding of how these investigations are carried out, to answer the questions.

Pooja designed her investigation to test this hypothesis: *The greater the concentration of copper added to cress seeds, the lower will be the percentage of seeds that germinate.*

Pooja took 40 Petri dishes containing cotton wool. She spread out 30 cress seeds onto the cotton wool in each dish. She then added to each dish 50 cm^3 of either water or one of three different concentrations of copper sulfate solution.

Pooja placed the dishes in a cabinet at a constant temperature of 25 °C. After three days she counted the number of seeds that had germinated in each dish.

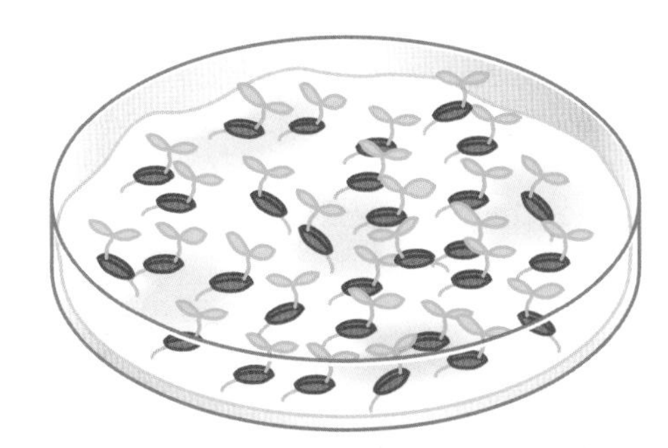

30 seeds on cotton wool. 50 cm^3 of water or copper sulfate solution added to each dish.

	Solution added to dish			
	Water	**0.001% copper sulfate**	**0.01% copper sulfate**	**0.1% copper sulfate**
Number of seeds that germinated	20	20	16	0
	19	16	3	0
	21	19	14	1
	17	15	12	7
	22	19	10	14
	24	23	8	0
	19	14	15	3
	25	17	6	1
	20	19	9	0
	22	18	11	1
Mean	20.9	18.1		2.7

A Pooja's results.

Answer the questions that follow.

1 Pooja chose to investigate whether the concentration of copper affected the germination of cress seeds. What kind of variable is the number of germinated seeds?

a a categoric variable
b a control variable
c an independent variable
d a discrete variable (*1 mark*)

2 Calculate the mean number of seeds that germinated in the 0.01% copper sulphate solution. (*2 marks*)

3 State the range for the number of seeds germinated in 0.001% copper sulfate solution. (*2 marks*)

4 ✎ Pooja decided that the differences between her results would be easier to see if she presented them as a graph. Explain whether she should use a bar chart or a line graph and say what she should plot on each axis. Quality of written communication is important in this answer. (*4 marks*)

5 Do Pooja's results agree with her hypothesis? Explain your answer. (*1 mark*)

6 Pooja showed her results to Tom. He said that her investigation might not be a 'fair test' because she had used copper sulfate solution. Look back at Pooja's hypothesis and explain what Tom meant. (*1 mark*)

7 Tom suggested that they repeat Pooja's investigation, using copper nitrate solution instead of copper sulfate. How would this prove whether her data was valid? (*2 marks*)

8 Describe **one** way in which Pooja's evidence could lead to society understanding more about pollution. (*1 mark*)

Total: 14 marks

Glossary

acid rain Acidic rain forms when sulfur dioxide dissolves in rainwater.
adaptation Characteristic that helps an organism to survive well in the environment.
adult cell cloning Cell fusion using an adult cell nucleus, as used to make Dolly the cloned sheep.
ancestor An organism that lived earlier and evolved into the descendant.
asexual Reproduction from one cell of an organism, without fusion of male and female sex cells.
bacterium (plural **bacteria**) A type of single-celled microorganism without a nucleus.
biodiversity The amount of variety in living species on the Earth.
callus Cluster of cells grown by tissue culture.
carbon dioxide store Something that locks carbon dioxide away for a long time, such as a tree or fossil fuel.
cell fusion Inserting a nucleus from one cell into an unfertilised egg that has had its nucleus removed.
characteristic Some feature of an organism, such as its size or colour.
chromosome Long molecule of DNA found in the nucleus of a cell.
clone Individual with the same genes as another.
cutting Small piece of plant, such as leaf or stem, used to grow new plants.
deforestation Cutting down large areas of trees, such as rainforest.
DNA Chemical that makes up genes and chromosomes.
egg Female sex cell.
embryo Ball of cells made by dividing fertilised egg cell.
embryo splitting Separating cells of an embryo at an early stage so that they will grow into separate individuals that are clones.
environmental factor One part of the environment, such as light or warmth, or another animal or plant.
evolve Change in characteristics over time.
evolutionary tree Diagram of ancestors and descendants, showing which organisms evolved into which.
extinct A species which has no individuals still living.
fertilise When the male sex cell and female sex cell fuse.
fertiliser Chemical put on crops to help them grow better.
fuse Join together.
gene Small piece of DNA that contains the code for a characteristic.
genetic engineering Taking genes from one organism and putting them into the cells of another organism so that the cells have the characteristic of the new gene.
genetically modified An organism that has been genetically engineered.
global warming Increasing temperatures on the Earth, thought to be due to increasing greenhouse gases.
greenhouse effect The trapping of warmth by the Earth's atmosphere that keeps the surface of the Earth warm enough for life.
greenhouse gases Gases that trap the warmth of the Earth, such as carbon dioxide and methane.
herbicide Chemical that kills plants, used to treat weeds in crops.
indicator Organism that shows if an environment is polluted.
inherit Get from a parent.
landfill Tipping refuse into large holes in the ground.
methane A greenhouse gas that is released from decaying plants, particularly in cows' guts and from flooded paddy fields.
mutation A change in a gene.
natural selection When the environment selects individuals so that only the best adapted survive and breed.
nitrogen oxides Gases that pollute the air, formed from burning fossil fuels.
nucleus Structure in cells that contains the chromosomes.
pesticide Chemical used to kill pests, such as insects on crops.
raw material Materials that can be made into other things, such as ore, crude oil, tree trunks.
reproduce Make more of.
selective breeding When a breeder chooses individuals with particular characteristics to breed from.
sewage Waste from houses and offices.
sexual reproduction Producing offspring by the fusion of male and female sex cells.
species (fossil) A group of organisms that have the same characteristics; (living) A group of organisms that can interbreed successfully.
standard of living The quality of life, including how easy it is to get food and to do all the things you want to do easily and safely.
sulfur dioxide The acidic gas formed when sulfur burns in air.
sustainable Something that can be done forever without damaging the environment.
tissue culture Making many new plants from cells taken from the parent plant.
toxic Poisonous.
transgenic organism An organism that contains genes from another organism.
variation Differences between organisms.

Products from rocks

A The construction of Terminal 5 at Heathrow.

B The main terminal building contains 80 000 tonnes of steel.

Building the new terminal at Heathrow involves using lots of different materials. From the concrete foundations to the steel in the roof, from the copper wiring to the heating system, all of these have been designed and made for a purpose. They are all made from substances that come out of the ground, either as rocks or as crude oil. Useful substances are then extracted from these rocks or oil, and turned into materials to make the building and everything in it.

By the end of this unit you should:

- be able to describe how rocks like limestone provide essential building materials
- be able to explain the difference between elements, compounds and mixtures, and how the elements are arranged in the Periodic Table
- know how to balance equations
- be able to describe how metals are very useful in our everyday lives
- be able to explain how metals can be extracted from rocks according to their reactivity
- be able to evaluate how extracting metals from rocks impacts on the environment
- be able to describe how crude oil can be separated to provide fuels.

1 Look at this list of materials used to build Heathrow. Think about the raw materials they are made from.
steel girders; concrete surrounds; cement between bricks; glass lights; copper wiring; arrival board screen; plastic seats; chrome plating; heating fuel; petrol for cars; concrete for the runways; tar on the roads leading to the terminal

a Sort them into groups.
b Explain why you have grouped them in the way that you have.

V

D

Digging up rocks

By the end of this topic you should be able to:

- explain the differences between rocks, ores and minerals
- describe how limestone quarrying can affect people and their environment.

Building the new terminal at Heathrow is a big project. It needs a lot of materials to build all the different parts. Many of the materials have come from **limestone**.

The Earth's crust is made of **rocks**. These can be tiny, such as grains of sand, or enormous boulders weighing many tonnes. The chemical compounds and elements that are found in rocks are called **minerals**. An **ore** is a rock that contains enough mineral to make it worth spending money to **extract** it. Common ores include limestone (which contains calcium carbonate), rock salt (containing sodium chloride) and bauxite (aluminium oxide). All of these ores must be quarried before the minerals in them can be extracted and used.

1 What is the difference between a rock, a mineral and an ore?

2 **a** Name three minerals.
b Name three ores.

A Limestone quarrying.

iron pyrites

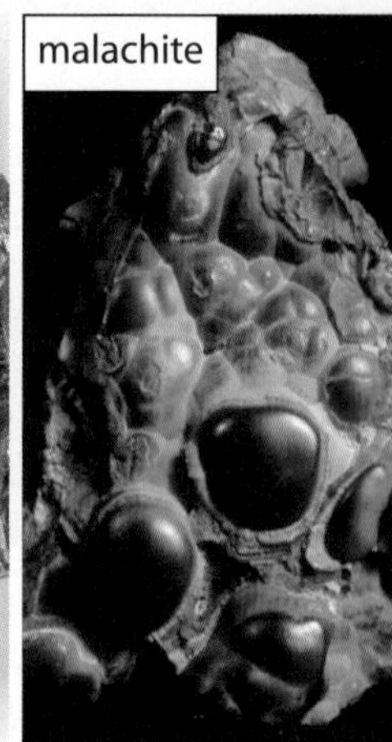

malachite

haematite

B These are all minerals.

Impact of quarrying

The problem with quarrying is that it affects everything around it. Before a new quarry is built, the owners have to consider a lot of different factors. These include:

- the cost of extracting the mineral from the ore
- the quality of the ore
- the availability of people living nearby to work in the quarry
- the amount of noise that will result from blasting away the rock
- the dust and dirt that will be produced
- the impact of quarrying on the landscape
- the impact of the quarry on the local wildlife
- the extra traffic
- the effect on local shops
- the effect on tourism if the quarry is in a tourist area
- the need for new roads to transport the ore away
- the price that the minerals could be sold for
- what could be done with the quarry after all the ore has been removed.

3 Sort out the factors in the list on the left into those that affect the environment and those that do not. Put them in a table.

4 What effect would a large quarry have on:
a the local shops
b tourism?

5 Who would make money from having a quarry? Explain your answer.

Some rocks do not need to be purified because they can be used as important raw materials in their impure state. Limestone is an example. It can be used as aggregate. Aggregate is the base for roads before the tarmac surface is laid on top. Limestone is also turned into other chemicals such as glass and cement. These are mainly used in the construction industry. It is also used in agriculture to neutralise acidic soils.

C Many people think that this is a waste of limestone, as other rocks would be just as good.

D Lime is used to neutralise acidic soils.

Rock salt is used directly on the ground to prevent roads freezing over. It is also used to make sodium hydroxide, chlorine and hydrogen which are all used in industry. **Sand** (silicon dioxide) is used for making glass.

E Rock salt lowers the melting point of ice.

6 a Name three rocks that can be used without being purified.
b How are these rocks used?

7 Make a table to show the advantages and disadvantages of building a limestone quarry.

Using limestone for buildings

By the end of this topic you should be able to:

- describe how limestone can be used for building
- describe the effect of this on people and the environment
- describe how limestone can be used to make cement, concrete, mortar, slaked lime and glass.

Limestone is a very important material for the building trade. In limestone areas like Yorkshire it is used instead of bricks to build the walls of houses. This is because it is cheap and available. It can also be used to make cement, mortar, concrete, slaked lime and glass.

1 Name two things in photograph A that are not made from limestone or one of its products.

A Mortar is used to stick stones or bricks together, blocks of limestone are used instead of bricks, concrete is used for paths and glass is used in windows.

Cement is made by heating powdered limestone and clay together in a rotary kiln. The kiln is a hot oven which rotates so that the limestone and clay mix together. The heat makes the two react together to form cement.

Cement can be mixed with sand and water to form **mortar**. This wet mixture is used to stick bricks or slabs of stone together. As it dries it hardens. **Concrete** is made by mixing cement with water, sand and crushed rock. As this mixture sets, it forms a hard, stone-like solid. **Reinforced concrete** is concrete that is allowed to set around a steel support. The steel support makes the concrete much stronger.

B Reinforced concrete will make this building much stronger.

2 What is the difference between cement, mortar and concrete?

3 How is the concrete in the Terminal 5 building at Heathrow made strong?

Glass is made by heating limestone with sodium carbonate and pure sand. At first the mixture melts, but when it cools down it forms a transparent solid.

Limestone is heated to make cement and glass. Limestone is mainly **calcium carbonate** ($CaCO_3$). When it is heated it **decomposes** (breaks down) to form calcium oxide and carbon dioxide. Calcium oxide is also known as **quicklime.** When calcium oxide reacts with water it forms calcium hydroxide. Calcium hydroxide is also known as **slaked lime**.

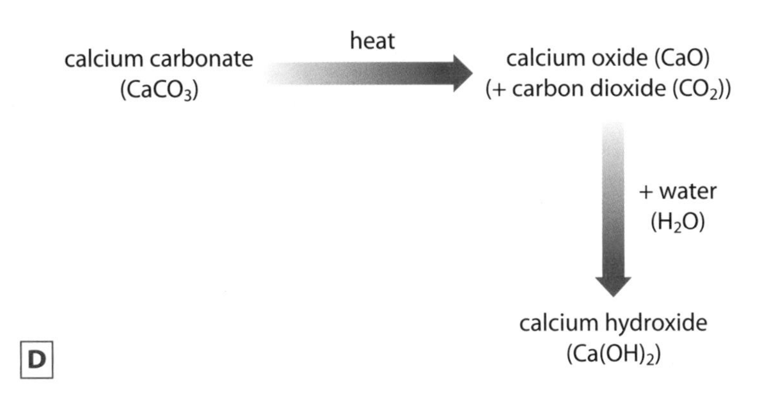

D

Weathering of buildings

The problem with using cement, mortar or limestone in buildings, is that they gradually wear away. This is because carbon dioxide dissolved in rainwater reacts with them. This often leaves gaps between bricks. These gaps need to be filled in (pointed) with new mortar or cement. Acid rain, formed by air pollution, speeds up the wearing away of cement, mortar and limestone.

E Acid rain will gradually wear away this brickwork.

6 Why aren't many limestone houses built today?

C The colours in this glass are made by adding certain metal compounds.

4 Write word equations using chemical names, to show what happens:
  a when calcium carbonate is heated
  b when water is added to calcium oxide.

5 Why do you think that these compounds have a chemical name (e.g. calcium oxide) and a non-chemical name (e.g. quicklime)?

7 Draw a concept map to show how limestone is turned into six other materials.

Atoms and elements

By the end of this topic you should be able to:

- recall that all substances are made up of atoms
- describe how elements contain only one type of atom
- explain what is inside atoms
- describe how elements can be shown as symbols.

A John Dalton was born in 1766.

The word '**atom**' was first used in the fifth century BC by a Greek thinker called Leucippus. He thought that everything was made up of tiny particles that could not be divided. However, nobody believed him, and his ideas were ignored until John Dalton put forward his **atomic theory** in 1807. The main points of Dalton's theory are:

- All matter is made up of tiny particles called atoms.
- Atoms cannot be created or destroyed.
- All of the atoms of an element are exactly alike.
- Atoms of one element are completely different to the atoms of another element.
- Atoms combine in small numbers to form molecules.

1 What points did Dalton make about atoms that Leucippus didn't?

We now know that Dalton's theory is inaccurate, but it was the basis of scientific research in the 19th century.

Atoms are in fact made up of three different kinds of particles: **protons**, **neutrons** and **electrons**. The masses of protons and neutrons are tiny, (0.000 000 000 000 000 000 000 00167 g each!) so we say that they each have a mass of 1 unit. Electrons are even smaller and weigh 1/2000 of this! Protons have a positive electric charge and neutrons are neutral. The protons and neutrons are found in the central **nucleus** of an atom. The negatively charged electrons are found moving around outside the nucleus. Atoms are always neutral, so the number of protons and electrons must always be the same.

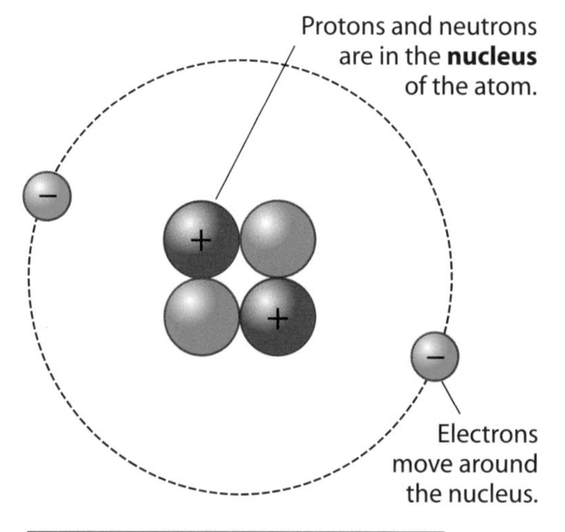

B The structure of a helium atom.

2 What are atoms made up of?

3 How does our idea of atoms today differ from Dalton's idea?

4 Copy and complete table C.

Particle	Mass (units)	Charge
Proton		
Neutron		
Electron		

C

- An **element** is a substance that only contains one kind of atom. Each atom in an element has the same number of protons. Different elements have different numbers of protons in their atoms.
- The **atomic number** of an element is its number of protons.
- The **mass number** of an element is the number of protons plus the number of neutrons.

For example, carbon has six protons and six neutrons. So its atomic number is 6 and its mass number is 12.

D All of these different elements have their own atomic number and mass number.

5 Copy and complete table E.

Element	Number of protons	Number of neutrons	Atomic number	Mass number
Fluorine	9	10		
Potassium	19	20		
Iron	26	30		

E

Each element also has its own **symbol**. This is a shorthand way of writing down the element, and is the same whatever the language spoken by scientists. The symbols have one or two letters. For example, oxygen is O and sodium is Na. The first letter is always a capital letter and the second one is lower case. It is important that you remember this. For example, Co is cobalt metal which is an element, not CO which is carbon monoxide gas and a compound.

6 Why do we use symbols for elements?

7 Draw the nucleus of a nitrogen atom, which has an atomic number of 7 and a mass number of 14.

P S P D P P

Elements in the Periodic Table

By the end of this topic you should be able to:

- recall that there are about 100 different elements
- describe how the Periodic Table shows all of these elements
- explain how the Periodic Table tells us about the properties of elements.

Nearly every chemistry lab has a **Periodic Table** hanging on the wall.

1	2	transition metals										3	4	5	6	7	0
H hydrogen																	He helium
Li lithium	Be beryllium											B boron	C carbon	N nitrogen	O oxygen	F fluorine	Ne neon
Na sodium	Mg magnesium											Al aluminium	Si silicon	P phosphorus	S sulphur	Cl chlorine	Ar argon
K potassium	Ca calcium	Sc scandium	Ti titanium	V vanadium	Cr chromium	Mn manganese	Fe iron	Co cobalt	Ni nickel	Cu copper	Zn zinc	Ga gallium	Ge germanium	As arsenic	Se selenium	Br bromine	Kr krypton
Rb rubidium	Sr strontium	Y yttrium	Zr zirconium	Nb niobium	Mo molybdenum	Tc technetium	Ru ruthenium	Rh rhodium	Pd palladium	Ag silver	Cd cadmium	In indium	Sn tin	Sb antimony	Te tellurium	I iodine	Xe xenon
Cs caesium	Ba barium	La lanthanum	Hf hafnium	Ta tantalum	W tungsten	Re rhenium	Os osmium	Ir iridium	Pt platinum	Au gold	Hg mercury	Tl thallium	Pb lead	Bi bismuth	Po polonium	At astatine	Rn radon
Fr francium	Ra radium	Ac actinium															

A One of the most useful tools for a chemist.

Although it looks complicated, you do not have to learn it by heart, but do need to know how to use it. It is very useful in helping us to predict the properties of elements.

First, notice that it shows the symbol, mass number and atomic number for every element.

1 Use the Periodic Table to answer these questions:
a How many protons, neutrons and electrons are there in sodium?
b What are the symbols for:
(i) calcium **(ii)** copper?
c What is the atomic number of magnesium (third row)?
d What is the mass number of krypton (fourth row)?
e How many protons and neutrons are there in:
(i) beryllium (second row) **(ii)** arsenic (fourth row)?

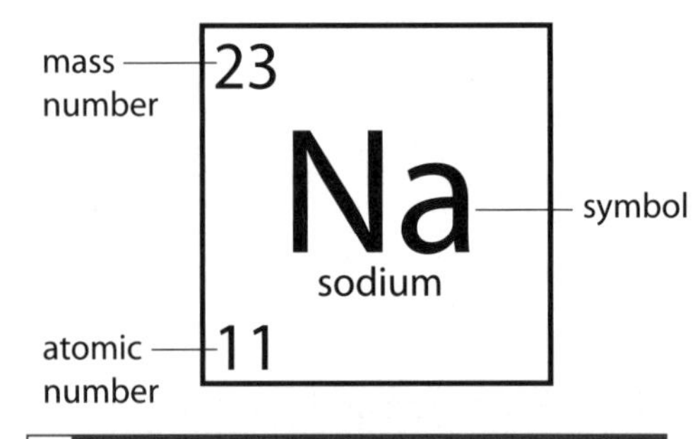

B This tells us all about a sodium atom.

All of the elements are arranged in increasing atomic number in the Periodic Table. You will also see a black line towards the right-hand side. This divides metals like iron on the left-hand side from non-metals like nitrogen to the right.

The vertical columns are called **groups**. There are eight different groups, called Groups 1–8. There are also a lot of metals between Groups 2 and 3. These are called the transition metals. All of the elements in a particular group have similar properties. For example, the elements in Group 1 are all very reactive metals. They have to be stored under oil as they react very quickly with air and water.

The non-metals in Group 7 are also similar to one another. They are all typical non-metals, and do not conduct heat or electricity.

2 There are no blanks in the Periodic Table. What does that tell you?

3 How many non-metals are there in the Periodic Table?

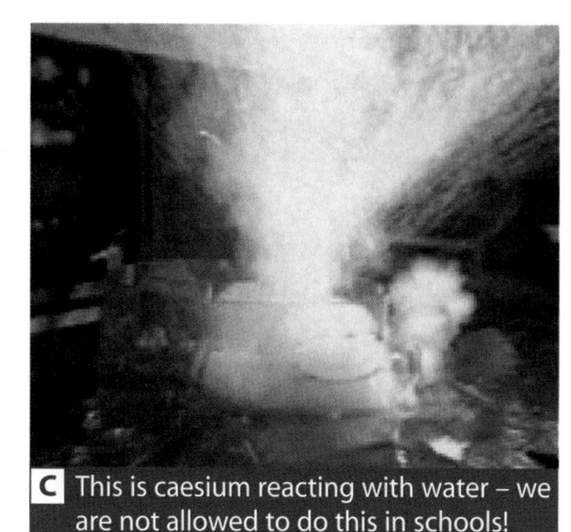

C This is caesium reacting with water – we are not allowed to do this in schools!

D These are typical non-metals.

4 Do Group 7 elements tend to have high or low melting points? (*Hint*: look at the photographs in D.)

5 Draw up a table to show similarities and differences between the elements in Groups 1 and 7.

Similarly, all of the elements in Group 0 are very unreactive gases, mostly used in lighting. Helium is also used to fill balloons as it is lighter than air, and unlike hydrogen, will not explode in air.

E The Hindenburg exploded as it was full of hydrogen which caught fire.

6 Look at argon and calcium in the Periodic Table.
a How are they similar?
b Explain, in terms of atomic particles, why this happens.

7 Write an article for the school magazine, explaining why the Periodic Table is so important.

P P P D P P

Compounds and formulae

By the end of this topic you should be able to:

- describe how elements react to form compounds
- explain how the formula of a compound shows the number and type of element in it
- explain how electrons are either shared or transferred between atoms to form compounds.

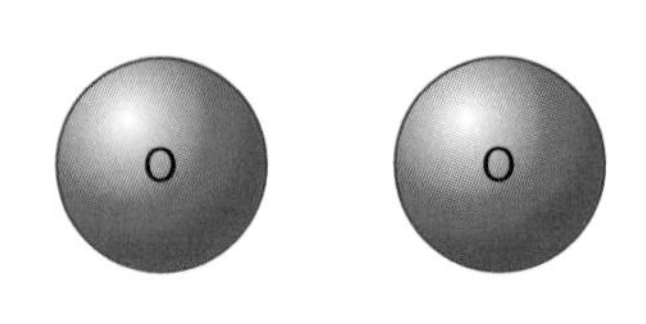

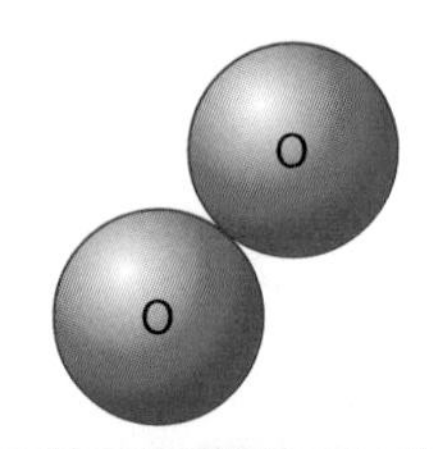

A Two oxygen atoms join to make one oxygen molecule.

Atoms do not normally exist by themselves. Instead many of them join together to form **molecules**. A molecule may contain two or more atoms which are the same. For example, oxygen gas contains two atoms per molecule. These two atoms are held together by chemical bonds.

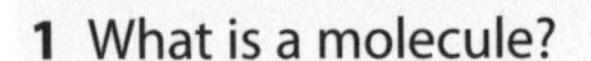

1 What is a molecule?

When two or more *different* kinds of atoms join together they form a **compound**. A compound can often look completely different to the elements that it is made from.

B Table salt is made from a very reactive metal and a toxic gas.

2 What is a compound?

3 Hydrogen and oxygen gases combine to form water. How does the appearance of water differ from that of hydrogen and oxygen?

Compounds are formed when atoms give, take or share electrons. A metal usually *gives* one or more electrons to a non-metal. For example, sodium gives chlorine an electron when the two react together.

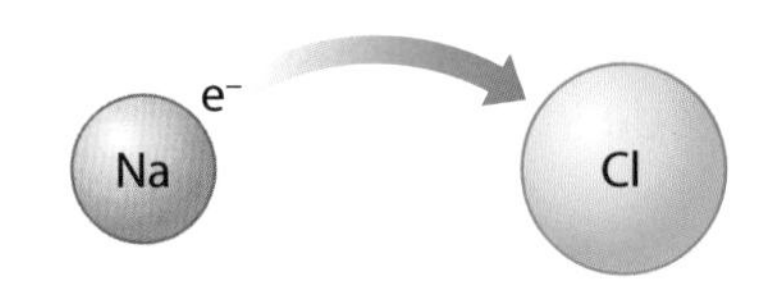

C Sodium loses an electron and chlorine gains one.

Sometimes atoms of different elements don't give or take electrons, but share them. These elements are usually non-metals. For example, in water an oxygen atom shares an electron with each of two hydrogen atoms.

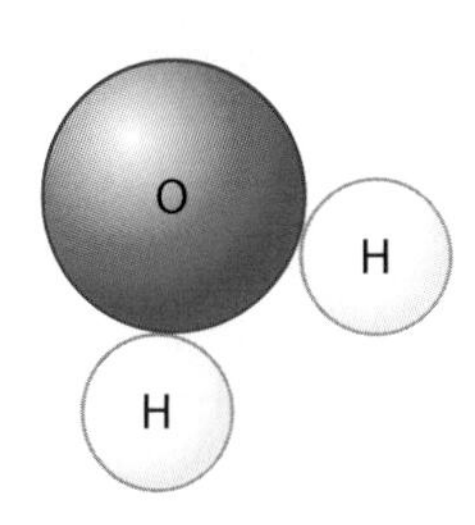

D A water molecule.

4 Name a compound where electrons are shared.

5 Name an atom which can take an electron from another atom.

We use a **formula** to show what is in a molecule of either an element or a compound. Oxygen molecules have two atoms of oxygen, so we write O_2. The '2' shows that there are two atoms of oxygen in each molecule of oxygen. The formula of a compound shows us the number and type of atoms that are joined together to form the compound. The formula of sodium chloride is NaCl. This tells us that one sodium atom is joined with one chlorine atom. We do not write Na_1Cl_1 as we do not bother with 1s in formulae.

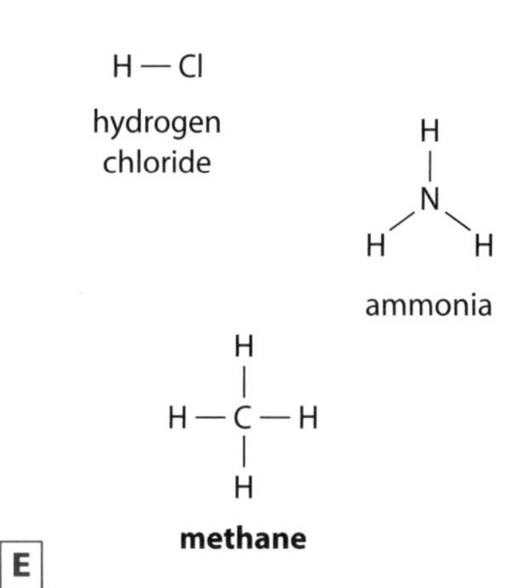

E

6 a How many atoms of hydrogen and chlorine are there in each molecule of hydrogen chloride?
b Write the formula for hydrogen chloride.

7 a Find the number of atoms of each element in ammonia and methane.
b Now write the formulae of ammonia and methane.

8 Limestone (calcium carbonate) has the formula $CaCO_3$.
a What type of atoms does it contain?
b How many of each type of atom are there?

9 Copy and complete table F. You can check the symbols using the Periodic Table.

Molecule	Formula	Atoms of each element present
Chlorine	Cl_2	
Carbon dioxide	CO_2	
Ethane	C_2H_6	
Sulfuric acid	H_2SO_4	

F

10 Write the formula of each of the following. Write the metal bit first:
a copper oxide – with one atom of copper and one atom of oxygen
b sulfur dioxide – with one sulfur atom and two oxygen atoms
c nitric acid – with one hydrogen, one nitrogen and three oxygen atoms.

11 a Write a definition for each of the words below.
b Give examples of each.

atom	formula
compound	molecule
element	symbol

Writing equations

By the end of this topic you should be able to:

- write and balance symbol equations
- use them to show what happens when you heat metal carbonates
- describe how atoms are not lost or gained during chemical reactions.

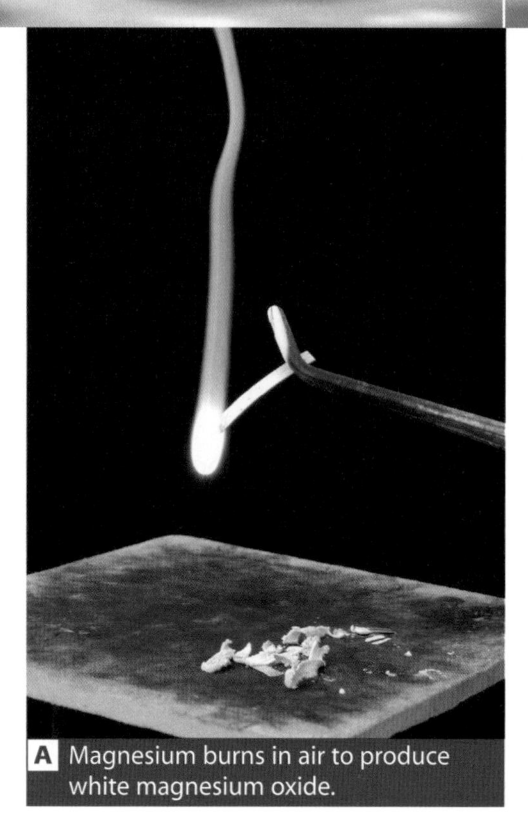

A Magnesium burns in air to produce white magnesium oxide.

We can show what happens in this reaction using a **word equation**:

magnesium + oxygen → magnesium oxide
reactants → product

Atoms are not gained or lost in a chemical reaction so the mass of the products is the same as the mass of the reactants. A **balanced symbol equation** shows this.

For example the equation for burning magnesium is:

$$2Mg(s) + O_2(g) \rightarrow 2MgO(s)$$

Balanced symbol equations are more useful as they show you:

- the formulae of the reactants and products
- the numbers of atoms and molecules taking part.

They can also be used to calculate the amounts of substances reacting or produced in a reaction.

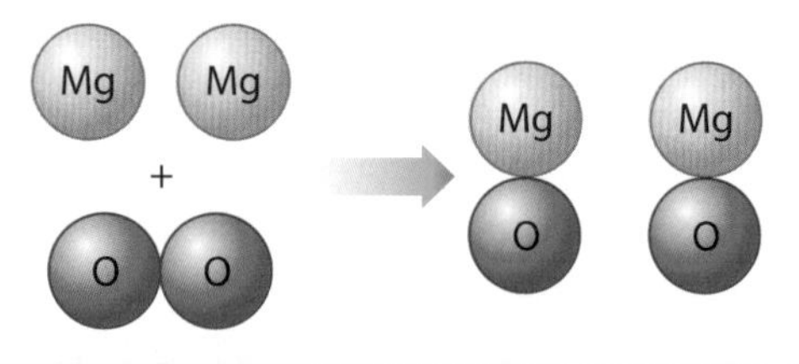

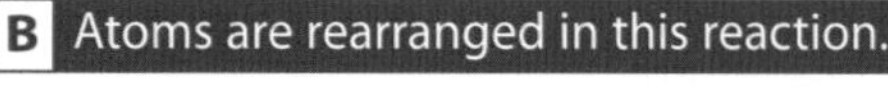

B Atoms are rearranged in this reaction.

State symbols are written in brackets after the symbols. They tell you the physical state of each substance:

- gas (g)
- liquid (l)
- solid (s)
- dissolved in water (aq)

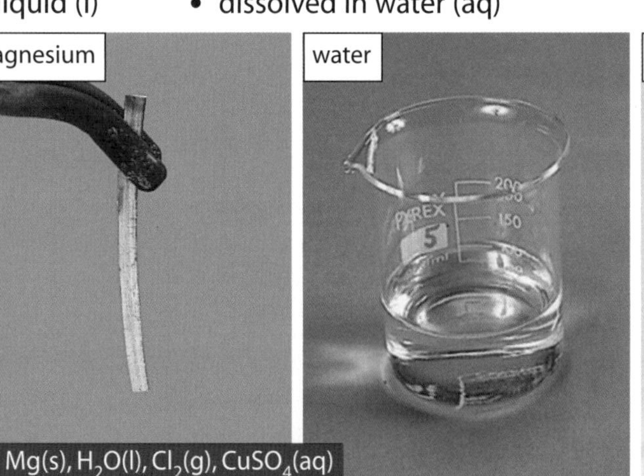

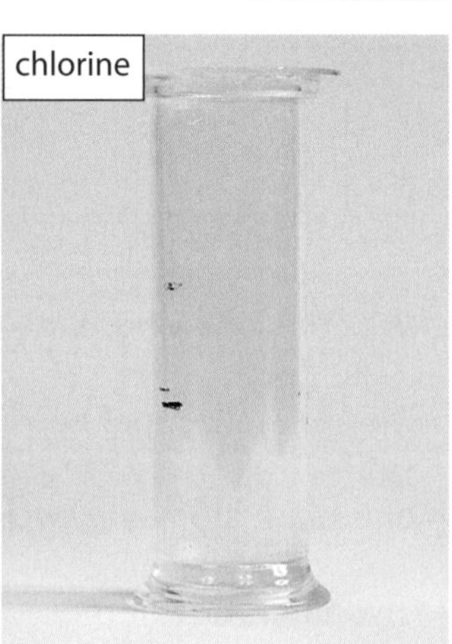

C $Mg(s)$, $H_2O(l)$, $Cl_2(g)$, $CuSO_4(aq)$

1 a What is the formula for magnesium oxide?
b Is magnesium oxide a solid, a liquid or a gas? How would you show this in the formula?

However, symbol equations don't tell you how fast the reaction is, or what you would see. Burning magnesium gives out a bright white light and lots of energy, but the equation doesn't tell you this.

You can balance an equation by counting the number of atoms of each element on each side of the equation. For example, hydrogen (H_2) reacts with chlorine (Cl_2) to form hydrogen chloride (HCl):

hydrogen + chlorine → hydrogen chloride

$H_2(g) + Cl_2(g) \rightarrow HCl(g)$

Reactants	Products
H: 2	H: 1
Cl: 2	Cl: 1

There are not enough hydrogen and chlorine atoms on the right-hand side, so you double the number of hydrogen chloride molecules by writing a 2 in front of it.

$H_2(g) + Cl_2(g) \rightarrow 2HCl(g)$

Reactants	Products
H: 2	H: 2
Cl: 2	Cl: 2

There are now the same numbers of hydrogen and chlorine on each side of the equation. The equation is now balanced.

Remember that you can never change formulae when balancing equations.

2 Look at this equation.
$Mg(s) + 2HCl(aq) \rightarrow MgCl_2(aq) + H_2(g)$
a What are the reactants?
b What are the products?
c What does '(aq)' mean?
d Give two things you can predict that you would expect to see during this reaction.

P

3 Write balanced symbol equations for these reactions.
a hydrochloric acid + zinc → zinc chloride + hydrogen
$HCl(aq) + Zn(s) \rightarrow ZnCl_2(aq) + H_2(g)$
b sodium oxide + water → sodium hydroxide
$Na_2O(s) + H_2O(l) \rightarrow NaOH(aq)$

P

When calcium carbonate is heated to make quicklime it is broken up to form calcium oxide and carbon dioxide:

$CaCO_3(s) \rightarrow CaO(s) + CO_2(g)$

You will heat other metal carbonates, and find out what is formed. You will then be able to write equations for these reactions.

4 Write the equation for the reaction of calcium oxide with water to form calcium hydroxide ($Ca(OH)_2(aq)$). (*Hint*: the brackets mean that there are two oxygens and two hydrogens.)

P

D

5 Make a list of everything that this equation tells you:
$4Al(s) + 3O_2(g) \rightarrow 2Al_2O_3(s)$

P

P

Mining metals and their ores

By the end of this topic you should be able to:

- describe how some metals are more reactive than others
- describe how mining metal ores has a social, economic and environmental impact
- explain why we recycle metals.

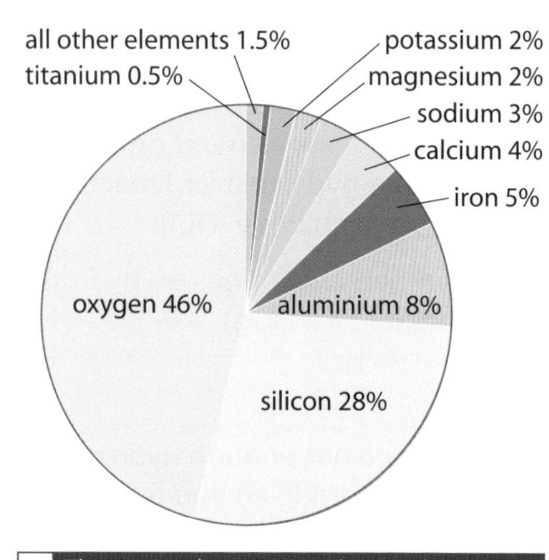

A The main elements in the Earth's crust.

The Earth's crust contains many different compounds as well as pure metals like gold.

1 Which element is the most common in the Earth's crust?

2 What are the two most common metals?

All of the other metals not shown in the pie chart are found in tiny amounts. These scarce metals include gold, silver and platinum. These are all **unreactive** and do not form compounds easily. This means that they are found as pure elements.

B Gold used to be mined in Wales. Small amounts are still mined in Wales today.

Other metals are more reactive. They form compounds more easily and are found in ores. Ores contain enough metal to make it economic to extract the metal. Whether it is economic changes over time as the amount of metal available or its price changes.

C Open-cast mining presents the same problems for the environment as limestone quarrying.

D A spoil tip.

There are four stages in extracting a metal from its ore.

- The ore is mined by quarrying, tunnelling or open-cast digging.
- The ore is separated from any impurities. These impurities produce a lot of waste. Mining waste is called spoil. The spoil is left by the mine and looks very ugly. Spoil tips may contain large amounts of poisonous metals like copper and lead. These metals dissolve in rainwater and leak into the soil. Any plants that are sown on spoil tips die.
- The ore is converted to the metal. This may be done by heating the metal ore in air or with carbon. In both cases impurities can react with air to form poisonous substances like sulfur dioxide.
- The metal formed is then purified.

Pure metals are turned into useful products. For example, electrical wires, saucepans, bikes and so on. When the product has worn out and is no longer useful, it can be recycled. This means reprocessing the metal and reusing it. Recycling is particularly important for metals like aluminium that are expensive to make. It is also important for very rare metals.

Recycling metals:

- saves money
- means that reserves of metal in the ground last longer
- avoids waste
- avoids the effects of mining on people and the environment
- saves energy.

3 A new important use has been found for a very rare metal.
 a What will be the probable effect on the price of the metal?
 b Will this make mining the metal more worthwhile?

4 What are the similarities between open-cast mining and quarrying for limestone?

5 How do spoil tips affect the environment?

6 Why do plants growing on spoil tips die?

E

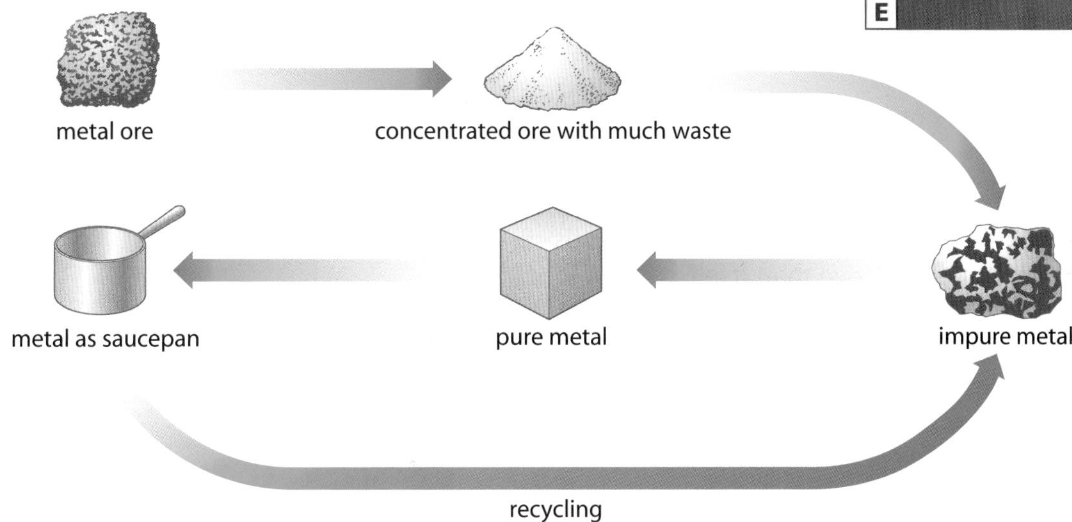

F Extraction and one use of aluminium.

7 What are the advantages of recycling?

8 Draw a flowchart showing how a copper kettle can be made and then recycled.

Extracting metals from their ores

By the end of this topic you should be able to:

- describe how metals that are less reactive than carbon are extracted from their ores by reduction with carbon
- explain how iron oxide is reduced to iron in a blast furnace
- describe how metals more reactive than carbon are extracted using electrolysis.

The **reactivity series** lists metals in order of their reactivity. Those at the top (like potassium) are the most **reactive** and those at the bottom (like platinum) are the least reactive.

potassium	Use electrolysis.	hard to extract
sodium		
lithium		
calcium		
magnesium		
aluminium		
(carbon)		
zinc	Ores are heated with carbon.	getting harder to extract
iron		
tin		
lead		
(hydrogen)		
copper	Found in native state (although copper is sometimes found as an ore).	easy to extract
silver		
gold		
platinum		

A The reactivity series.

1 a Name a metal more reactive than aluminium.
b Name a metal less reactive than tin.

A reactive metal will displace a less reactive metal from its compounds. For example, if you heat magnesium with copper oxide there is a violent reaction. Magnesium oxide and copper are formed:

magnesium + copper oxide → magnesium oxide + copper

$$Mg(s) + CuO(s) \rightarrow MgO(s) + Cu(s)$$

Carbon and hydrogen are included in the reactivity series as they are very useful for extracting metals from their ores.

Less reactive metals like iron and copper are usually found as oxides. If you heat either of these oxides with carbon, the carbon removes the oxygen. The reaction gives the metal and carbon dioxide. For example:

copper oxide + carbon → copper + carbon dioxide

$$CuO(s) + C(s) \rightarrow Cu(s) + CO_2(g)$$

This is because carbon is more reactive than either of these metals. This type of reaction is called **reduction**. Carbon is used to reduce oxides of metals less reactive than itself.

B Magnesium displaces copper.

2 Name three metals that can be extracted from their ores using carbon.

3 Write a balanced symbol equation to show the reaction between carbon and zinc oxide. Solid zinc oxide has the formula ZnO.

4 a Write a balanced symbol equation to show the reaction between magnesium and zinc oxide. Solid magnesium oxide (MgO) is formed.
b What would be the problem with getting zinc from this mixture?

Iron is a very important metal, used mainly as steel to make large items such as bridges, oil rigs and cars. Iron is found as iron oxide (Fe_2O_3) in an ore called **haematite**. It is mixed with coke (pure carbon) in a **blast furnace**. In this reaction the iron oxide is **reduced** to iron (it loses oxygen) and the carbon is **oxidised** to carbon dioxide. Because it gets so hot, the iron formed melts. This molten iron flows to the bottom of the furnace.

C Molten iron being tapped off (removed) from a blast furnace.

5 How many atoms of iron and oxygen are there in a molecule of iron oxide?

6 Write a word equation for the reaction in a blast furnace.

As we cannot use carbon to reduce the oxides of metals more reactive than carbon, we have to use another method. This process is called **electrolysis**. Electrolysis involves splitting up a compound using electricity. This is expensive, so it is only used when necessary and economically viable.

D Aluminium is formed by the electrolysis of bauxite. Bauxite is mainly aluminium oxide.

7 Name two other metals which can only be obtained by electrolysis.

8 Explain how the reactivity series is useful in telling you how to extract a metal from its ore.

P

P

D

P

P

Turning iron into steel

By the end of this topic you should be able to:

- explain that the iron coming out of a blast furnace is impure and has limited uses
- describe how removing these impurities produces pure iron
- describe how the atoms are arranged in pure iron
- describe how and why iron is turned into steel
- explain why the properties of alloys are related to their structures.

The iron that comes out of a blast furnace is only about 96% pure. It contains impurities like carbon, silicon, sulfur and phosphorus. These impurities make the iron brittle and so it is not very useful. To get pure iron you need to remove all of these impurities. This is done by reacting these elements with oxygen. The oxides formed are then easily separated from the iron.

Like all other metals, the atoms in pure iron are arranged in a regular pattern. However, the layers can slide over one another. This means that pure iron is soft and easily shaped.

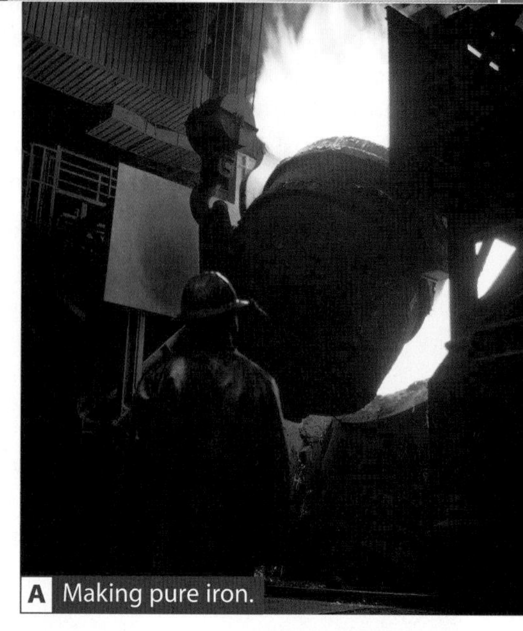

A Making pure iron.

1 Name the impurities found in iron.

2 How are they removed?

3 What effect do these impurities have on the iron?

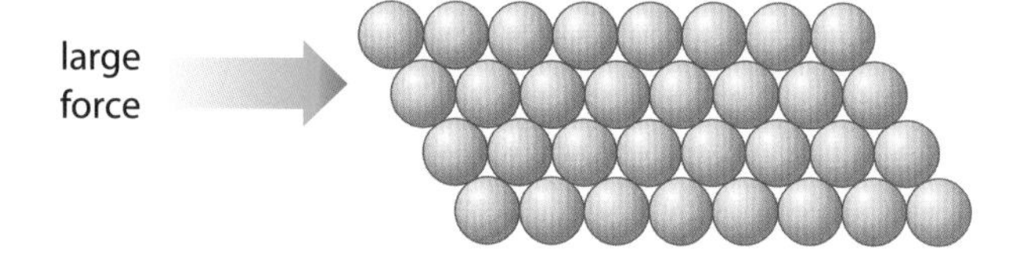

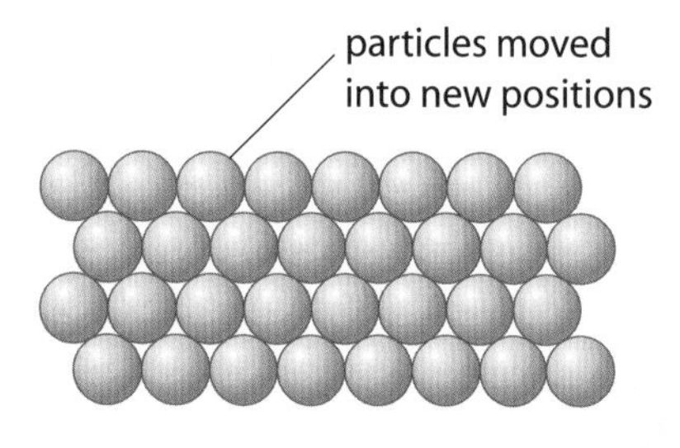

B The metal atoms slide over each other if a force is applied.

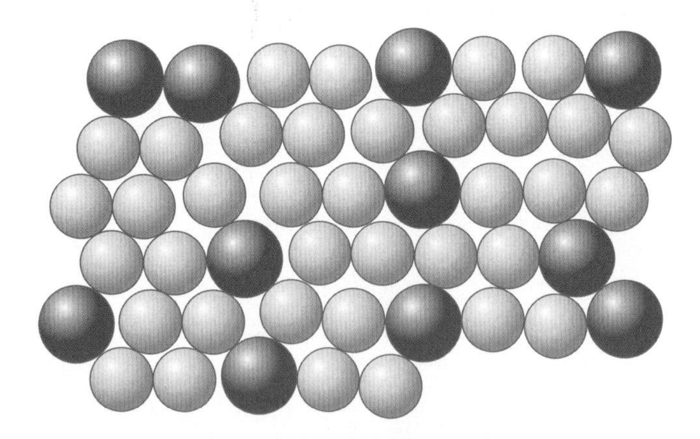

C In alloys the layers can't slide so much now.

Steel alloys

The softness of iron means that it is too soft for many uses. Most iron is converted into **steels**. Steels are **alloys**. Steel alloys are a mixture of iron with carbon or other metals. The different-sized atoms which are added distort the layers of the structure of the pure metal. This makes it more difficult for them to slide over one another, and so alloys are harder.

4 What is an alloy?

5 Why is pure iron no good for making the girders used in building Terminal 5 at Heathrow?

Alloys can be designed to have properties for specific uses. Most steels are made by adding carefully calculated amounts of carbon to the pure iron. **Low-carbon steels** are made by adding small amounts (up to 0.25%) of carbon to the molten pure iron. This small amount of carbon makes the iron harder and stronger. These steels are easily shaped, and are used for things like wire, nails and car bodies.

If slightly more carbon is added (up to 1.5%), **high-carbon steels** are made. These are harder and stronger, but are brittle. They are used for cutting tools and masonry nails. Masonry nails are used for hammering into bricks or concrete. Using high-carbon steel nails means that they do not bend as low-carbon iron nails would.

D These girders are made of low-carbon steel.

E Masonry nails are made from high-carbon steel.

Stainless steels are alloys of iron with chromium and nickel. They make the iron resistant to corrosion. These alloys are used for things like cutlery, but are very expensive.

F Stainless steel cutlery will not rust.

6 What is the difference in the properties of low-carbon steel and high-carbon steel?

7 Suggest another item that needs to be made out of stainless steel.

8 Write an article for a school encyclopaedia on 'steel'. It should be no longer than 100 words.

P D P P

Transition metals and alloys

By the end of this topic you should be able to:

- explain why the transition metals are useful structural materials
- list the names and uses of some common alloys
- describe the benefits and drawbacks of using metals for building structures
- explain what smart alloys are.

The **transition metals** are found in the centre of the Periodic Table between Groups 2 and 3. They include a range of different elements, but they tend to have similar properties.

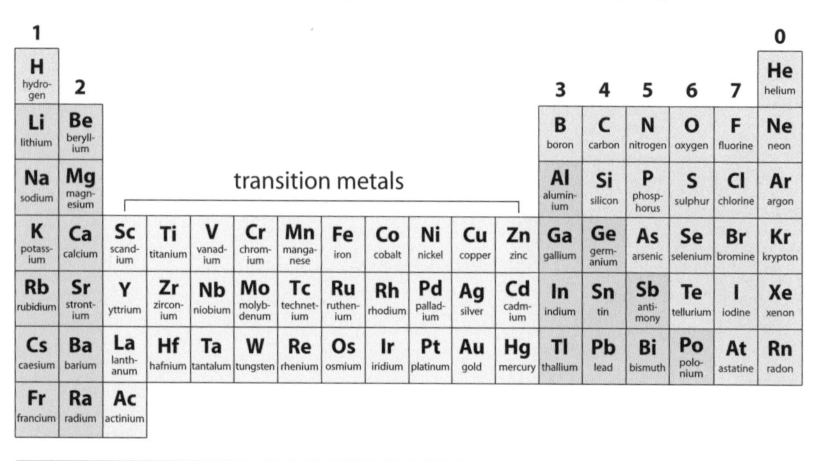

1	2	transition metals										3	4	5	6	7	0
H hydrogen																	He helium
Li lithium	Be beryllium											B boron	C carbon	N nitrogen	O oxygen	F fluorine	Ne neon
Na sodium	Mg magnesium											Al aluminium	Si silicon	P phosphorus	S sulphur	Cl chlorine	Ar argon
K potassium	Ca calcium	Sc scandium	Ti titanium	V vanadium	Cr chromium	Mn manganese	Fe iron	Co cobalt	Ni nickel	Cu copper	Zn zinc	Ga gallium	Ge germanium	As arsenic	Se selenium	Br bromine	Kr krypton
Rb rubidium	Sr strontium	Y yttrium	Zr zirconium	Nb niobium	Mo molybdenum	Tc technetium	Ru ruthenium	Rh rhodium	Pd palladium	Ag silver	Cd cadmium	In indium	Sn tin	Sb antimony	Te tellurium	I iodine	Xe xenon
Cs caesium	Ba barium	La lanthanum	Hf hafnium	Ta tantalum	W tungsten	Re rhenium	Os osmium	Ir iridium	Pt platinum	Au gold	Hg mercury	Tl thallium	Pb lead	Bi bismuth	Po polonium	At astatine	Rn radon
Fr francium	Ra radium	Ac actinium															

A Many of the best known transition metals are found in the top row.

1 Look at diagram A. Are gold and platinum transition metals?

The transition metals are all typical metals:

- They have high melting and boiling points.
- They are good conductors of heat and electricity.
- They are hard, tough and generally strong, although they can be easily bent or hammered into shape.

However, the transition metals are also very different to the metals found in Group 1. They do not react rapidly with water or air, although some of them corrode slowly. They are therefore useful as structural materials and for making things that conduct heat or electricity easily. Tungsten (symbol W) with its very high melting point is used for filaments in light bulbs.

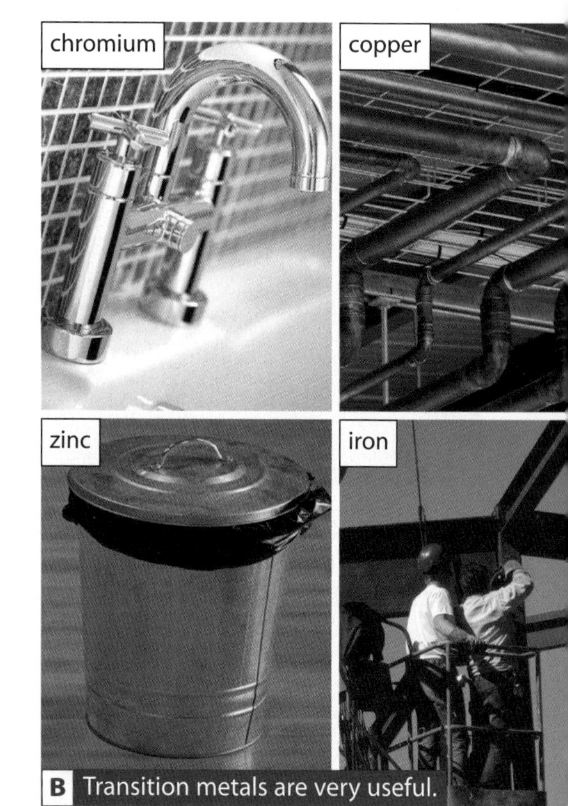

B Transition metals are very useful.

2 Why is iron used so much for making things like bridges, oil rigs and cars?

3 Why is copper used for electrical wiring?

4 a Write down three differences between Group 1 metals and transition metals.

b Explain how these differences make transition metals useful for:

(i) filaments in light bulbs

(ii) scissor blades.

Transition metals are even more useful when they are mixed with other metals to form alloys. Pure copper and gold are too soft for many uses, so they are mixed with small amounts of other metals to make them harder. Although aluminium is not a transition metal, it also has important structural properties as it has such a low density. It is therefore also turned into important alloys for use in aircraft.

Name of alloy	Composition	Special property	Uses
Brass	70% copper, 30% zinc	harder than pure copper	electrical fittings, screws
Bronze	90% copper, 10% tin	harder than pure copper	bells
Cupro-nickel	75% copper, 25% nickel	harder than pure copper	coins
Duralumin	96% aluminium, 4% copper	stronger than pure aluminium and lighter than copper	aircraft

C Properties and uses of some alloys.

D 'Copper' coins are 97% copper, 2.5% zinc and 0.5% tin.

5 Why are copper and gold turned into alloys?

6 Cupro-nickel is used for 'silver' coins. What is the difference in composition between this and the alloy used for 'copper' coins?

Smart alloys

Recently **smart alloys** have been developed. These are more correctly called shape-memory alloys, because they return to their original shape after they have been deformed. The two main types are the copper–zinc–aluminium alloys and the more expensive nickel–titanium alloys. They can be used in construction, and more and more in medicine. For example, dental braces have been developed which maintain a constant pressure on the teeth.

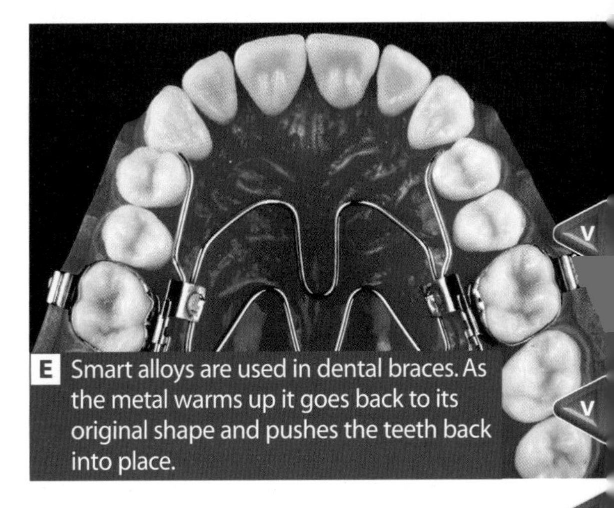

E Smart alloys are used in dental braces. As the metal warms up it goes back to its original shape and pushes the teeth back into place.

7 Why are smart alloys so useful? Give an example.

8 Suggest why nickel–titanium alloy is more expensive than the copper–zinc–aluminium one?

9 Is an alloy a compound or a mixture? Explain your answer, giving examples.

Copper, titanium and aluminium

By the end of this topic you should be able to:

- explain why copper is useful for electrical wiring and plumbing
- explain how copper is extracted from its ore, and its impact on the environment
- explain why new ways of extracting copper from low-grade ores are being researched
- explain why titanium and aluminium are also such useful metals but extracting them is difficult and expensive.

Copper is a very useful metal because it is an excellent conductor of heat and electricity. It has been used for centuries in such things as saucepans and kettles as it conducts heat so well. It is used for water pipes and radiators. (Although plumbers prefer using copper, plastic pipes are also used now.) Copper is valuable in electrical wiring as it is such a good electrical conductor.

1 Why is copper used in wiring?

2 Why was copper used in kettles?

A People collect old copper kettles.

B Copper must be 99.95% pure to be a good electrical conductor.

Copper extraction

Copper is found mainly as the ore chalcopyrite ($CuFeS_2$). This ore only contains about 0.5–2.0% copper, but copper is so valuable that it is still worth extracting.

The impurities are removed so the ore is more concentrated. The ore is then heated and turns into copper oxide. The copper oxide is reduced to copper using carbon. However, the whole extraction process produces a lot of waste which is often dumped in rivers.

New ways of extracting copper from other ores which only contain small amounts of copper are being researched. This limits the damage to the environment. These methods include using bacteria which convert the copper to copper sulfate. Iron is then added to the copper sulfate. Copper is formed in a **displacement reaction**.

3 If 1000 kg (1 tonne) of ore produces less than 4 kg of pure copper, how much waste will be left?

4 Why can carbon reduce copper oxide to copper?

5 Write a word equation for the reaction of iron with copper sulfate.

Titanium is another very valuable transition metal. It has some very useful properties:

- It is very strong.
- It does not corrode.
- It can withstand very high temperatures.
- It is considerably less dense than iron (lighter for its size).

6 Look at the photographs in C. Why is titanium so useful in:
 a aeroplanes
 b racing bicycles?

C Titanium's properties make it very useful.

Titanium is found as titanium dioxide (TiO_2) in an ore called rutile. It is extracted in two stages:

- The titanium dioxide reacts with chlorine (Cl_2) to form titanium chloride ($TiCl_4$).
- The titanium chloride is then reacted with sodium. This reaction is done in an inert atmosphere of argon. Pure titanium and sodium chloride (NaCl) are formed.

7 **a** For the extraction of titanium, write:
 (i) a word equation
 (ii) a balanced symbol equation for the second reaction only.
 b Why is argon used in this reaction?
 c Why is this method more expensive than reacting titanium oxide with carbon?

Aluminium is not a transition metal, but it has some useful properties:

- It has a very low density.
- It is very strong.
- It is covered with a very thin layer of aluminium oxide, which is insoluble in water and prevents the aluminium from corroding.

D Aluminium has some very useful properties. It is used to make furniture, drinks cans and take away cartons.

Aluminium is found in an ore called bauxite. It is extracted using electricity, which is expensive and uses lots of energy.

8 Why is aluminium used in vehicles that need to be light?

9 What could we use instead of metals if we run out of them?

Mixtures

By the end of this topic you should be able to:

- explain what a mixture is
- describe how mixtures can be separated.

A Not only is there a mixture of different foods, each food is a mixture of different substances.

A compound is a substance that contains two or more elements chemically combined together. You have seen that alloys are **mixtures** of metals. A mixture is two or more substances which are *not* chemically combined together. These mixtures may be:

- two or more elements (for example, an alloy is usually two or more metals mixed together in the right proportions)
- two or more compounds (for example, sea water is a mixture of water and sodium chloride (common salt))
- an element and a compound (for example, air is a mixture of oxygen, nitrogen, the Group 0 gases, carbon dioxide and water vapour).

1 What is the difference between a compound and a mixture?

2 What are the formulae of water and sodium chloride?

3 Fish need oxygen dissolved in water so that they can breathe. Is this a mixture of two elements, an element and a compound, or two compounds?

Unlike compounds, it can be quite easy to separate a mixture into its various parts. This is because the chemical properties of the substances in the mixture are not changed.

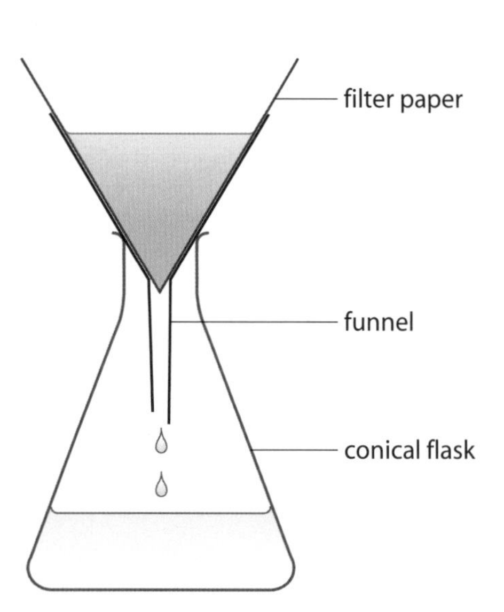

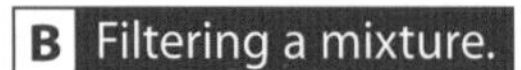

B Filtering a mixture.

C Obtaining salt from sea water.

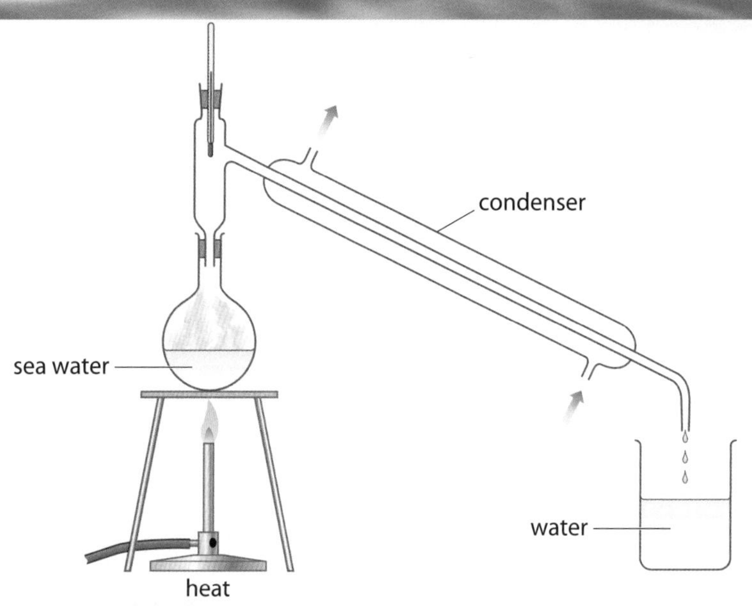

D The pure water evaporates.

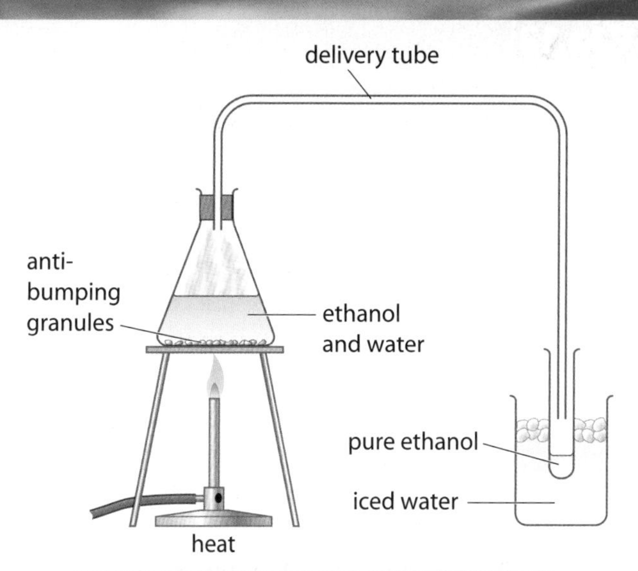

E Some alcoholic drinks are made by fractional distillation.

There are several common methods of separating mixtures:

- **Filtration** is used to separate an insoluble solid from a liquid. For example sand from water (see diagram B).
- **Evaporation** (see photograph C) separates a solid from a solution and saves the solid. A solution is made up of a solute (the solid) dissolved in a solvent (the liquid). The solvent is evaporated off, either by leaving the solution in a warm place or by heating it, until just the solid is left. For example, salt can be separated from sea water by allowing the water to evaporate in the Sun.
- **Distillation** (see diagram D) separates the solvent from a solution and saves the solvent. The mixture is heated, the solvent evaporates, and is then condensed and collected. For example, in very hot countries where there is not enough fresh water, drinking water can be obtained by distilling sea water.
- **Fractional distillation** (see photograph E) is used when two or more liquids are mixed together. The mixture is heated, and the liquid with the lowest boiling point evaporates first and is then condensed. Each liquid boils at a different temperature, so they can be separated.

4 Why do people use coffee filters?

5 Whisky is made by distilling a mixture of water (boiling point 100 °C) and ethanol (alcohol) (boiling point 78 °C).
  a What method of separation would you use?
  b Which would boil at the lower temperature: water or ethanol?
  c Wine is 14% ethanol and 86% water. How is the alcohol content of brandy increased to 40%?

6 How would you separate a mixture of sugar and glass?

7 Iron is magnetic. Why is iron easier to recycle than other metals?

8 'Mixtures are easily separated.' Explain this statement. Give examples in your answer.

P
D
P
P

Useful substances from crude oil

By the end of this topic you should be able to:

- explain that crude oil is a mixture of many compounds
- describe how to separate them by fractional distillation
- explain that each fraction has different uses.

Crude oil was made millions of years ago. Plants and animals living in the sea died and fell to the seabed. Here they were covered in sediment, which in time became rock. Pressure from these rocks and heat from the Earth turned these remains into oil and natural gas. As all living things contain many carbon compounds, the oil made from them is also rich in carbon compounds. Crude oil is pumped from under the ground or sea. It is a thick black mixture, containing thousands of different compounds.

The compounds in crude oil have boiling points between room temperature (20 °C) and about 400 °C. This means that you can separate them using fractional distillation, which depends on the liquids in the mixture having different boiling points. This is done in a huge fractionating column.

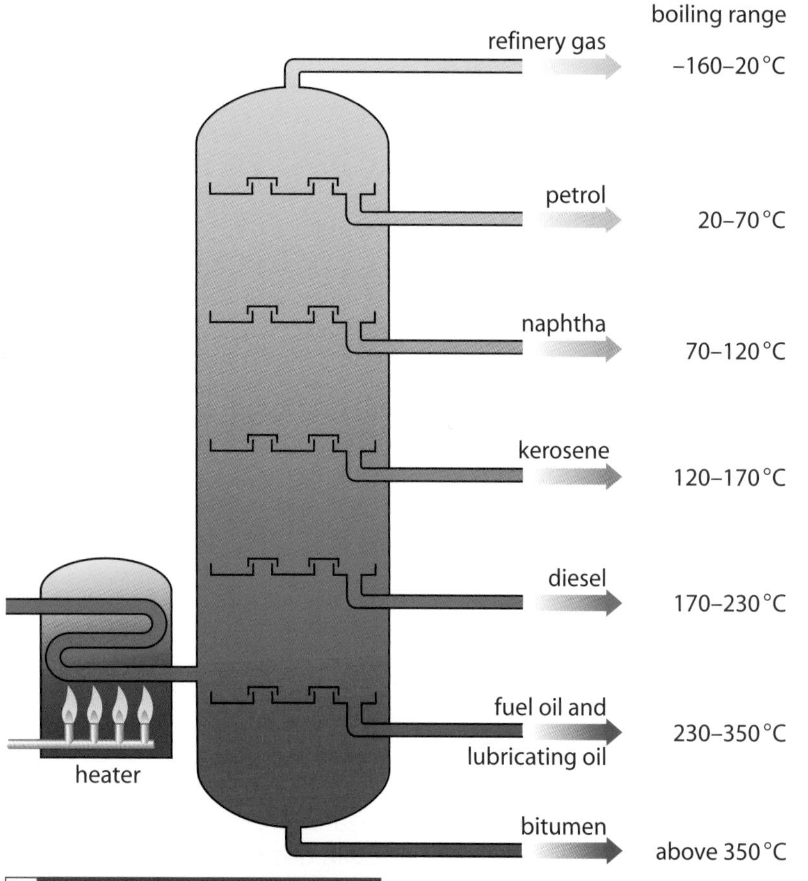

B Inside a fractionating column.

A The North Sea has provided us with oil for years.

1 How was crude oil formed?

2 Why is there so much carbon in oil?

C Oil refineries have enormous fractionating columns like this.

The crude oil is heated to turn it into a vapour. The vapour then passes up the tower, which is at different temperatures. When a compound is cool enough it turns back into its liquid state; that is it condenses. A liquid boils at the same temperature as its vapour condenses. So a fraction that boils at a certain temperature will condense at the same temperature. The crude oil separates into different fractions with different boiling points, and each fraction is run off separately.

The fractions formed all have their own special uses which are described in table D.

Fraction	Name	Uses
Below 20 °C	gas	fuel for gas ovens, chemicals
20–70 °C	petrol	petrol for vehicles, chemicals
70–120 °C	naphtha	chemicals
120–170 °C	kerosene	fuel for heating, jet engines, chemicals
170–230 °C	diesel	fuel for diesel engines
230–350 °C	fuel oil and lubricating oil	fuel for ships, waxes, lubricants
350 °C and above	bitumen	roofing, tar for roads

D Fractions of crude oil and their uses.

3 Look at diagram C.
a How does the temperature change going up the column?
b How many fractions are separated out?
c Is naphtha a gas or a liquid at 160 °C?

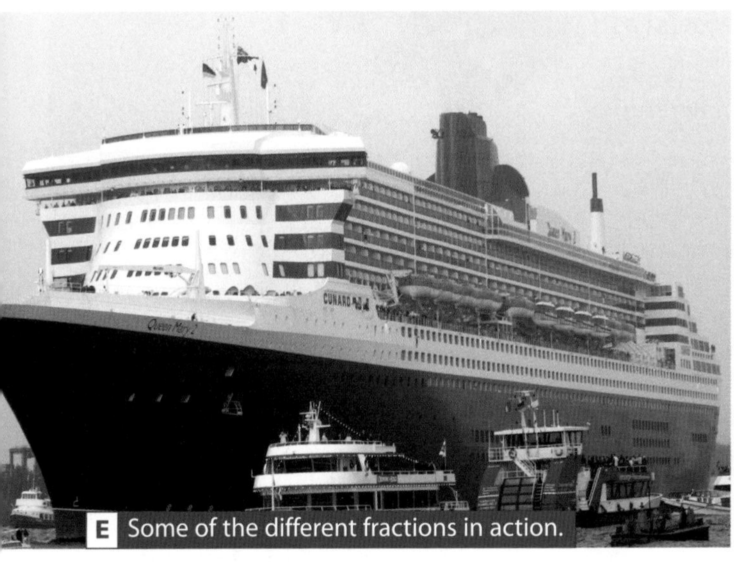

E Some of the different fractions in action.

4 Look at the photographs in E.
a What fuel does a ship use?
b What is in the tar being spread on the road?

5 What do all of the fractions that boil below 170 °C have in common?

6 Which fraction is used for gas ovens?

7 How do cars that run on petrol and those that run on diesel differ? (*Hint*: use table D to help you.)

8 Why is crude oil so important?

Alkanes

By the end of this topic you should be able to:

- explain that the fractions obtained after the fractional distillation of oil contain molecules with similar numbers of carbon atoms
- describe what an alkane is
- give the general formula for an alkane and how alkane molecules can be represented
- explain that alkanes are found in crude oil.

Most of the compounds in crude oil consist of molecules that contain only hydrogen and carbon. These are called **hydrocarbons**. The hydrocarbon molecules in crude oil vary in size. The larger molecules contain the most carbon atoms. These larger molecules all:

- have the highest boiling points
- turn into vapours the least easily
- are the most viscous – they flow the least easily
- are difficult to set fire to and are the least flammable
- have dark colours.

This means that the large hydrocarbons are not very useful as they are not good fuels.

The smaller hydrocarbons with fewer carbon atoms have the opposite properties to the large ones:

- They have low boiling points.
- They turn into vapours easily.
- They are runny.
- They burn well.
- They are light in colour.

This makes them good fuels.

C Oil like this will be used to heat Heathrow Terminal 5.

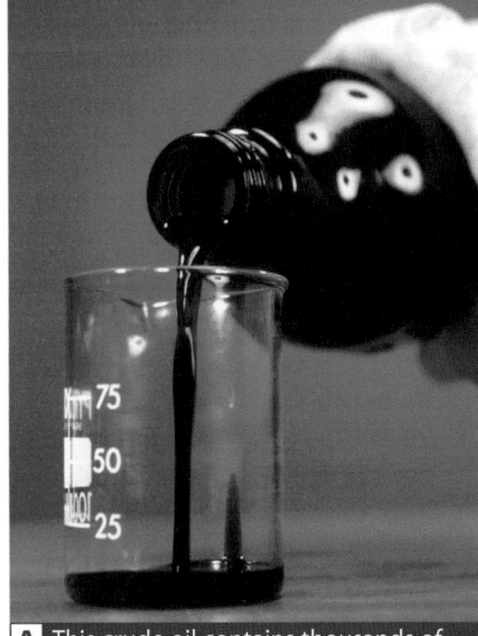

A This crude oil contains thousands of different hydrocarbons.

B Wax candles contain large hydrocarbons.

Look back at the last topic to help you answer these questions.

1. Which fraction from crude oil has the smallest molecules?
2. Which has bigger molecules, diesel or kerosene?
3. Why do you think that bitumen is made of very large hydrocarbon molecules?
4. What size molecules are found in wax?
5. What is a hydrocarbon?

Most of the molecules in crude oil are hydrocarbons called **alkanes**. The smallest alkane is **methane**, CH_4, which is found in the gas that we use in ovens and Bunsen burners. More alkanes are shown in diagram D.

ethane C_2H_6

propane C_3H_8

butane C_4H_{10}

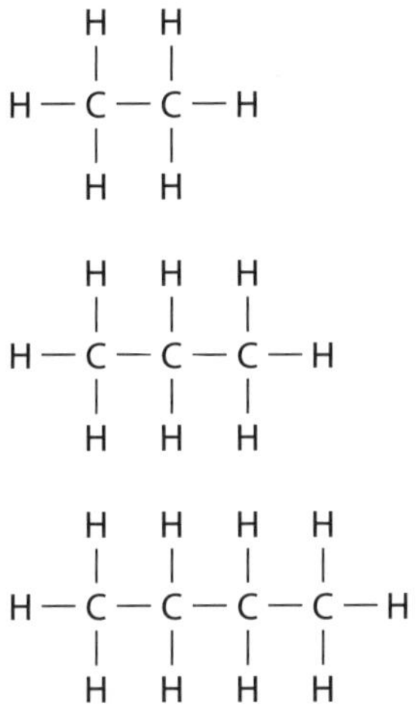

D These are the smallest alkanes, and are all gases at room temperature.

If you look at the formulae of these four alkanes, you will see a pattern. As we add one carbon atom to the molecule, we also add two hydrogen atoms. This means that alkanes have a general formula of C_nH_{2n+2} where n is the number of carbon atoms. Therefore pentane, which has five carbon atoms, has a formula C_5H_{12}, ($n = 5$). The alkanes are said to form a **homologous series**. This means that they all have similar properties, but the formula differs by CH_2 each time.

Alkanes are said to be **saturated**, as they cannot hold onto any more hydrogen atoms.

There is another way of writing butane. Instead of having all of the carbon atoms in a straight line, it can have a branch in the chain as shown in diagram E.

methyl propane butane

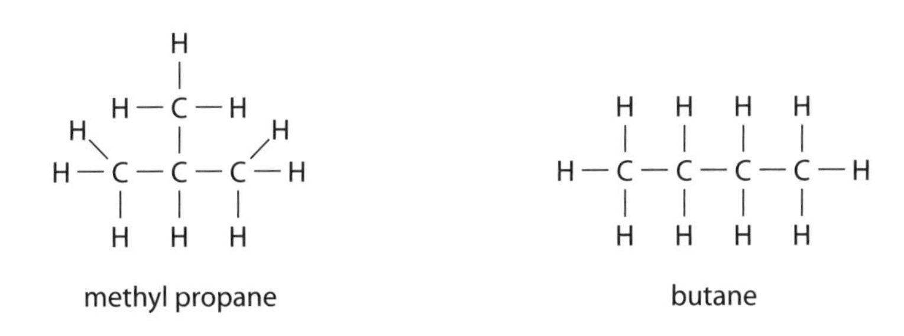

E These have the same formulae but different structures.

A totally different compound is formed, but it still has the same formula. This is called an isomer. The more carbon atoms there are in a hydrocarbon, the more isomers there are. This is why there are so many different compounds in crude oil – there are lots of isomers. $C_{10}H_{22}$ has 75 isomers, $C_{20}H_{42}$ has 366 319!

6 What is the formula for hexane, which has six carbon atoms?

7 What is a homologous series?

8 Draw a concept map showing everything that you know about oil.

Burning hydrocarbon fuels

By the end of this topic you should be able to:

- explain that most fuels contain carbon, hydrogen and possibly sulfur
- explain that the gases released into the atmosphere when a fuel burns may include water and carbon dioxide, which cause global warming, and sulfur dioxide which causes acid rain
- explain that particles which cause global dimming are also released.

Some of the fractions that are obtained from the fractional distillation of crude oil are used as fuels.

When these hydrocarbons burn they form water and carbon dioxide. For example, petrol contains octane (C_8H_{18}).

The equation for burning octane in air is:

octane + oxygen → carbon dioxide + water

$$C_8H_{18}(l) + 12.5O_2(g) \rightarrow 8CO_2(g) + 9H_2O(l)$$

If there is not enough oxygen present in a bus's engine, the fuel may not burn completely. This can leave carbon, which comes out of the exhaust as black particles. Most petrol is not pure hydrocarbon. It may also contain some sulfur. When sulfur burns, it forms sulfur dioxide. Sulfur dioxide is acidic when dissolved in water.

A A chemical reaction occurs inside the bus's engine.

1 What is formed when hydrocarbons burn completely?

2 What is also formed if there is not enough air?

3 What does sulfur burn to form?

The products of burning hydrocarbons can cause damage to the environment. More and more fossil fuels have been burned in the past century. This has contributed to an increase in the amount of **carbon dioxide** in the atmosphere. As a result of this, more of the Sun's heat is trapped in the atmosphere. This contributes to the **greenhouse effect**, and is linked to **global warming**.

4 What causes the greenhouse effect?

5 Why have more fossil fuels been burnt in the last century?

Rainwater is naturally slightly acidic. However, when **sulfur dioxide** dissolves in clouds it forms rain which is even more acidic. This is called **acid rain**. Acid rain can kill or harm plants and animals. It also attacks metals and limestone buildings.

The best way to avoid this is to remove sulfur from fuels before they are burnt. The sulfur which is removed can be used in chemical processes such as making sulfuric acid.

B The effects of acid rain.

C This sulfur will be shipped out of the harbour for use in the chemical industry.

Sulfur dioxide can also be removed from waste gases after burning. For example, filters that absorb sulfur dioxide are fitted into the chimneys of power stations.

6 What are the effects of acid rain?

7 Sulfur dioxide is acidic. What type of substance could be put into the power station chimneys to neutralise it?

When coal, oil and wood are burnt, they not only produce carbon dioxide and water. They also form tiny **particles** of soot, ash and sulfur compounds. These have been linked with **global dimming**. Global dimming decreases the amount of energy from the Sun entering the Earth's atmosphere. This may have a significant effect on the world's climate.

8 How could we try to reduce the amount of particles released from burning fossil fuels?

D Global dimming has been linked with monsoons in India.

9 Draw a table showing the causes and effects of carbon dioxide, sulfur dioxide and soot, and how they might be reduced.

P

P

D

P

P

Alternative fuels

By the end of this topic you should be able to:

- describe some developments in the production of better fuels
- evaluate them.

A All of these use crude oil as the raw material.

Oil and natural gas are essential substances. Not only are they very important sources of energy, they also provide the starting materials for a whole range of essential chemicals.

However, oil and other fossil fuels are non-renewable resources. This means that they are running out and cannot be replaced. Chart B shows how long they might last if we continue to use them at our current rate.

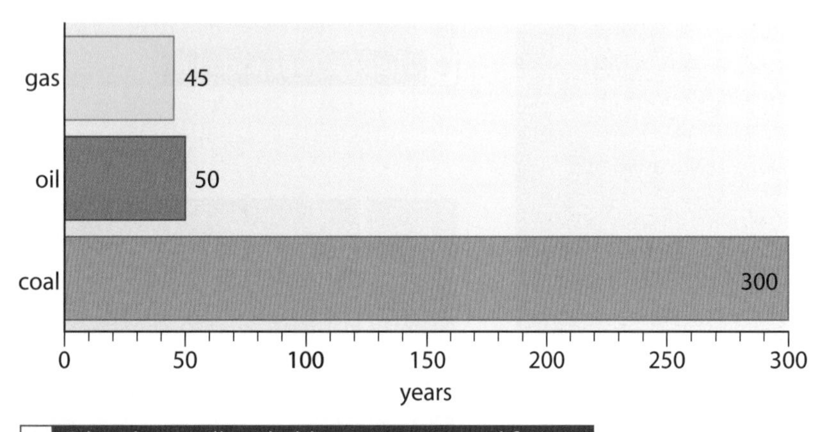

B Oil and gas will probably run out in your lifetime.

We therefore need to consider how to conserve these resources. We can do this by using them more efficiently and finding alternative fuels. We also need to find fuels that cause less pollution.

1 Which fuel will run out first?

2 How could we try to use less fuel?

3 What are the pollution problems caused by the fuels used at the moment?

Finding an alternative fuel has been partially solved by developing new fuels produced from living organisms, which are sustainable, or from renewable resources.

Biogas is a mixture of gases formed when bacteria break down plant material or animal waste products. These gases are flammable. However, not enough is made to provide enough energy for a large industrial plant. Scientists are trying to find out how to make greater amounts.

Ethanol may be used as a fuel in cars. It is produced in countries, such as Brazil, where sugarcane grows. Fermenting the sugarcane produces ethanol. In the United States maize is used. The starch in it is turned into sugars, which are then also fermented to make ethanol.

4 a Give one advantage and one disadvantage of biogas.
b In what parts of the world would biogas be a valuable fuel?

5 Why do you think ethanol is not used as a fuel in this country?

Another possible fuel is **hydrogen** gas. This has the advantage of only producing water when it burns. People who support hydrogen as a fuel argue that it is made from water which is a renewable resource. However, electricity is needed to split up the water into hydrogen and oxygen, and electricity is usually made by burning fossil fuels. Storing hydrogen is also a problem, especially if it is used to fuel cars. It is very difficult to turn hydrogen into a liquid, and a 50 litre tank (the norm for most cars) would only hold 4 grams of hydrogen gas. This much hydrogen would not produce very much energy when it burns, and the car would not travel very far!

Scientists have also researched electric cars. These run on batteries, and are, as such, non-polluting. However, again there is the problem of how the electricity is made to charge the batteries.

C Can this supply the fuel of the future?

D Electric cars run on rechargeable batteries.

6 a Give one advantage of using hydrogen as a fuel.
b Give two disadvantages.

7 a Why is using an electric car meant to cut down on air pollution?
b Why is this not always the case?

8 What alternatives are we trying to develop to replace hydrocarbon fuels?

Assessment exercises

Part A

1 Which of the following is NOT made from limestone?
- **a** cement
- **b** concrete
- **c** glass
- **d** plastic

(*1 mark*)

2 Which of the following is a compound?
- **a** carbon
- **b** oxygen
- **c** titanium
- **d** water

(*1 mark*)

3 Which of the following is NOT an alloy?
- **a** bronze
- **b** cupro-nickel
- **c** iron
- **d** steel

(*1 mark*)

4 The compounds found in crude oil are called…
- **a** carbohydrates.
- **b** carbon hydrides.
- **c** fractions.
- **d** hydrocarbons.

(*1 mark*)

5 Which of the following is NOT caused by burning fuels?
- **a** acid rain
- **b** global dimming
- **c** global warming
- **d** holes in the ozone layer

(*1 mark*)

6 The illustration below shows the mass and atomic numbers of Titanium.

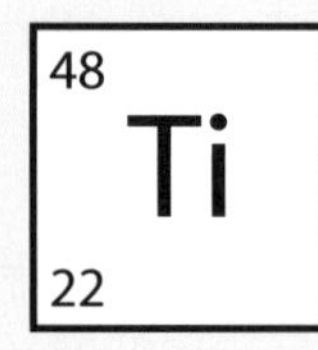

Use numbers listed (**a–d**) to answer questions **i–iv**.
- **a** 22
- **b** 26
- **c** 48
- **d** 70

- **i** What is the number of protons in titanium?
- **ii** What is the number of neutrons in titanium?
- **iii** How many electrons are there in titanium?
- **iv** What is the total of the number of protons and neutrons?

(*4 marks*)

7 The following are all alloys.
- **a** cupro-nickel
- **b** duralumin
- **c** low carbon steel
- **d** stainless steel

- **i** Which of these is used for coins?
- **ii** Which of these is used to make girders?
- **iii** Which of these is used in aircraft?
- **iv** Which of these is used for cutlery?

(*4 marks*)

8 From the periodic table shown below, choose the places where:
- **i** Metals that react quickly with water are found.
- **ii** Chlorine is found.
- **iii** Iron and copper are found.
- **iv** Gases like neon and argon are found.

(*4 marks*)

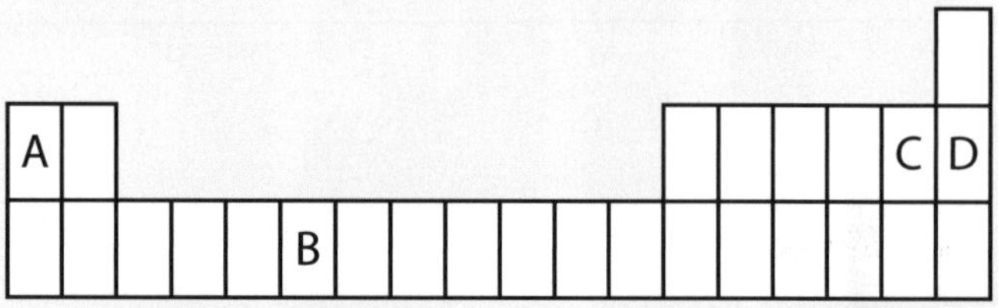

A
C D
B

Group 1 = A transition metals = B
Group 7 = C Group 0 = D

9 The flow chart below shows what happens when limestone is heated. Fill in blanks **i–iv** using words **a–d**.
- **a** calcium oxide
- **b** heat
- **c** slaked lime
- **d** water

(*4 marks*)

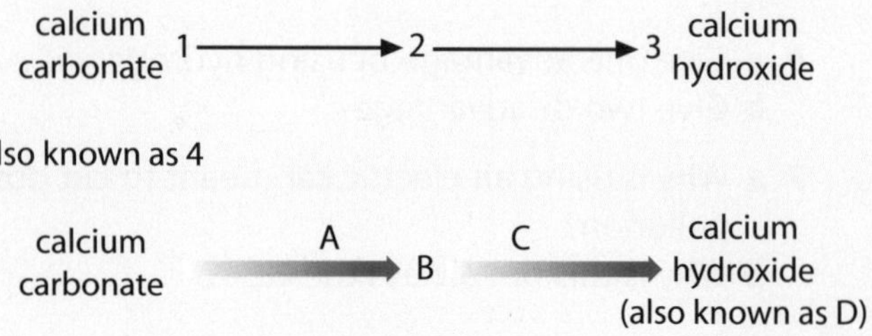

calcium carbonate 1 → 2 → 3 calcium hydroxide

also known as 4

calcium carbonate —A→ B —C→ calcium hydroxide (also known as D)

10 The diagram below shows a fractionating column used in industry. Choose from the diagram where the following happen:

i The largest molecules collect.
ii The crude oil is vapourised.
iii Fuels for vehicles are removed.
iv Gases leave the column.

(4 marks)

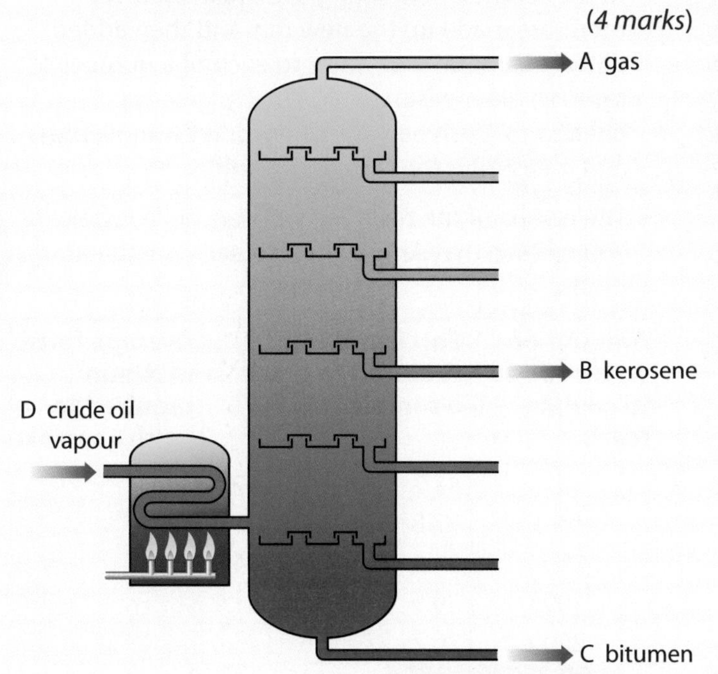

Total (Part A) 25 marks

Part B

1 A company wants to build a limestone quarry next to a small town.

a What is the chemical name for limestone? *(1 mark)*
b Name two building materials made from limestone. *(2 marks)*
c State one advantage to the company of having a quarry near to a town. *(1 mark)*
d Give two disadvantages to the residents of having a quarry near a town. *(2 marks)*
e How does mining a metal resemble limestone quarrying? *(2 marks)*

2 The formula of a compound shows us the number and type of atoms joined together in the compound.

a What is the formula of carbon dioxide? *(1 mark)*
b Iron oxide has the formula Fe_2O_3. How many iron atoms and oxygen atoms are there in one molecule of iron oxide? *(1 mark)*
c How many atoms are there in sulfuric acid H_2SO_4? *(1 mark)*

3 Look at the diagram below of the reactivity series.

Metal	
potassium	**hard to extract**
sodium	
lithium	
calcium	
magnesium	
aluminium	
(carbon)	
zinc	**getting harder to extract**
iron	
tin	
lead	
(hydrogen)	
copper	
silver	
gold	
platinum	**easy to extract**

a Name a metal that is found as an element in the ground. *(1 mark)*
b **i** Name a metal that cannot be extracted by reduction by carbon. *(1 mark)*
ii What is the name of the process by which this metal is extracted? *(1 mark)*
c Name a metal, other than iron, that could be reduced using carbon. *(1 mark)*

4 Titanium, copper and aluminium are all important metals.

a What special property makes copper so useful? *(1 mark)*
b Why is copper mixed with zinc to form brass? *(1 mark)*
c Copper is also alloyed with zinc and aluminium to form a smart alloy.
i What is a smart alloy? *(1 mark)*
ii Give one example of a use of a smart alloy. *(1 mark)*
d Why is titanium used in an artificial hip joint? *(2 marks)*

5 Alkanes have the general formula CnH$_2$n+2

a What is the formula of hexane which has 6 carbon atoms? *(1 mark)*
b Alkanes are hydrocarbons. What does the word 'hydrocarbon' mean? *(1 mark)*
c When hydrocarbon fuels burn they form carbon dioxide, water, sulfur dioxide and particles. What impact do **i)** sulfur dioxide and **ii)** particles have on the environment? *(2 marks)*

Total (Part B) 25 marks

Investigative Skills Assessment

The runways in airports have to be strong to withstand the pressure exerted when a plane lands. They are made of concrete, which is hard and strong. To find the best mix of concrete, Andrew made up four different mixes of sand, cement and gravel. Once the concrete sample had set he found out how much force was needed to crush the cement. He did this by placing the concrete sample in a G-clamp, and finding how many turns of the clamp were needed to crush the concrete.

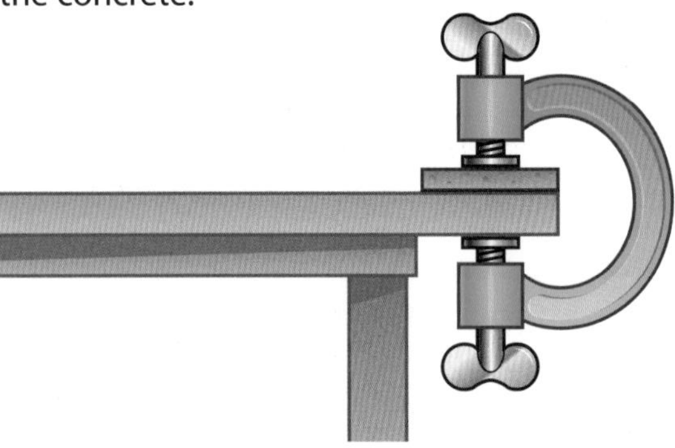

The results of Andrew's test are shown in Table A, below.

Ingredients	Number of spoonfuls of ingredient			
	Mix A	Mix B	Mix C	Mix D
Sand	3	5	2	2
Cement	3	2	5	2
Gravel	3	2	2	5
Number of turns of the G-clamp needed to crush the concrete	4	2	5	1

A Investigation results.

1 Starting with the strongest concrete first, list mixes A–D in order of decreasing strength. (*1 mark*)

2 Do these results fit any pattern? Explain your answer. (*2 marks*)

3 How could you try to improve the validity of the results? (*1 mark*)

4 Is there an independent variable in this experiment? If so, what is it? (*1 mark*)

Andrew decided that these results did not tell him enough about the correct mixture of concrete needed. He did some more research and found a recipe for a mixture that would give strong concrete. However, the recipe was a bit vague about the amount of water that should be added to the sand, cement and gravel mix. Andrew therefore made up the new mix and then added different amounts of water to each of a number of samples. He measured how much force was needed to crush each of the samples using a press and measuring the force in N/mm^2.

The results of the test are shown in Table B, below. You will see that Andrew tested each mixture four times.

Amount of water added to sample (cm^3)	Force in N/mm^2 required to crush a sample				Average force in N/mm^2 required to crush a sample
50	41.2	42.3	41.8	42.1	
100	37.8	37.3	37.5	37.6	
150	34.5	31.6	35.1	34.9	
200	30.9	31.1	34.2	31.0	
250	26.7	27.1	26.9	26.4	

B Investigation results.

5 Two of the results in Table B do not fit a pattern. Which results are they? (*1 mark*)

6 Complete Table B by working out the averages force required for each mixture. Omit the two results that do not fit a pattern. (*2 marks*)

7 What is the dependent variable in this experiment? (*1 mark*)

8 Why are these results more accurate than those shown in Table A? (*1 mark*)

9 What would be the best way of presenting these results? (*1 mark*)

10 Describe the pattern shown in these results. (*1 mark*)

11 Which set of results shows the greatest precision? (*1 mark*)

12 ✎ Would a more accurate force-meter have made much difference to the results? Explain your answer. Would you make any other changes to the experiment if you were to carry it out yourself? (*4 marks*)

Total = 15 marks

Glossary

acid rain Acidic rain forms when sulfur dioxide dissolves in rainwater.
alkanes Hydrocarbons with the general formula C_nH_{2n+2}.
alloy A mixture of metals in certain proportions.
aluminium A metal in Group 3 which is very strong but light.
atom The smallest part of an element.
atomic number The number of protons in the nucleus of an atom of an element.
atomic theory The theory that everything is made up of atoms.
balanced symbol equation When the number of different atoms in an equation are the same on both sides.
biogas Flammable gases formed when bacteria break down waste material.
blast furnace A large industrial furnace for extracting iron from iron ore.
calcium carbonate The chemical compound found in limestone ($CaCO_3$).
carbon dioxide Formed when hydrocarbons burn. An increase has led to global warming.
cement A substance made by heating limestone and clay together.
compound Contains two or more different elements.
concrete A substance made by mixing cement with water, sand and crushed rock.
condense Turning from a gas into a liquid.
copper A transition metal which is a very good conductor of heat and electricity.
crude oil Oil as it comes out of the ground. It is a mixture of hydrocarbons.
decompose Break down into simpler substances.
displacement reaction 'Competition' reaction between elements. A more reactive element displaces a less reactive one from its compound.
electrolysis Splitting up a compound using electricity.
electron A negatively charged particle which circles the nucleus of an atom.
element A substance with only one kind of atom.
ethanol A fuel made by the fermentation of sugar.
evaporation Used to separate a solid from a solution.
extract To get a mineral out of a rock.
filtration Used to separate a solid from a liquid.
formula Shows us what is in a molecule of an element or a compound.
fractional distillation Used to separate two or more liquids with different boiling points.
glass A substance made by heating limestone with sodium carbonate and sand.
global dimming The decrease in energy from the Sun reaching the Earth. The decrease is caused by the presence of particles in the atmosphere.
global warming Increasing temperatures on the Earth, thought to be due to increasing greenhouse gases.
greenhouse effect The trapping of warmth by the Earth's atmosphere that keeps the surface of the Earth warm enough for life.
group A vertical column in the Periodic Table.
haematite The main ore containing iron as iron oxide.
high carbon steel Iron plus slightly more (up to 1.5%) carbon added to it.
homologous series A group of similar compounds which just differ by a CH_2 group.
hydrocarbon Compound containing hydrogen and carbon only.
hydrogen A fuel made by splitting up water using electricity. It only makes water when it is burned.
limestone A rock used to make many different chemicals.
low carbon steel Iron plus small (up to 0.25%) amounts of carbon added to it.
mass number The number of protons plus the number of neutrons in an atom of an element.
methane The simplest alkane, formula CH_4.
mineral A chemical compound or element in a rock.
mixture Two or more substances mixed in any proportion which are not chemically combined.
molecule A pair or group of atoms bonded together.
mortar A substance made by mixing cement, sand and water.
neutron A neutral particle found in the nucleus of an atom of an element.
nucleus The centre of an atom where the protons and neutrons are found.
ore A rock with enough mineral to make it worth extracting.
oxidised Oxygen having been added.
particles Tiny particles of soot, ash and sulfur compounds formed when coal, oil and wood burn. They lead to global dimming.
Periodic Table A table showing all of the elements.
proton A positively charged particle found in the nucleus of an atom of an element.
quicklime Calcium oxide.
reactive A substance that readily reacts with water and air.
reactivity series A list of elements showing how reactive they are.
reduced Oxygen having been removed.
reduction The removal of oxygen from a compound.

reinforced concrete Concrete set around a steel support for strength.
rock What the Earth's crust is made of.
rock salt A rock that contains salt (sodium chloride).
sand A rock that contains silicon dioxide.
saturated Molecule containing no C=C double bonds.
slaked lime Calcium hydroxide.
smart alloys Alloys that return to their original shape after being deformed.
state symbols Tell us the physical state of substances.
steel Iron with carbon added to it to make the iron stronger.
sulfur dioxide The acidic gas formed when sulfur burns in air.
symbol A shorthand way of writing an element.
titanium A transition metal which is strong and will not corrode.
transition metals The metals in the centre of the Periodic Table between Groups 2 and 3.
unreactive Elements that do not react with other elements easily.
word equation A way of showing what happens in a chemical reaction.

Oils, earth and atmosphere

Many of us enjoy eating crisps. There is a lot of chemistry in a bag of crisps. The crisps are made from potatoes, fried using vegetable oils. Additives are also used to provide the different flavours and keep them fresh. You might think that the packet contains air, but the oxygen in air would make the crisps go off. Nitrogen is used instead. Then there is the bag which is made from poly(propene), a type of plastic. Plastic is made from crude oil which is formed inside the Earth. Once you have eaten your crisps you throw away the packet. What happens to the packet once you have thrown it away?

A There is a lot of chemistry in a bag of crisps.

By the end of this unit you should:

- know how crude oil can be used to make polymers
- be able to compare the different ways in which polymers can be disposed of
- be able to compare how ethanol can be made from renewable and non-renewable sources
- know how vegetable oils are used in our diet and their effect on health
- be able to evaluate the use of additives in foods
- be able to explain why the theory of continental drift took some time to be accepted
- understand why scientists cannot accurately predict when earthquakes and volcanic eruptions will occur
- know how the Earth's atmosphere has changed since the Earth was young
- be able to explain how human activities are changing the Earth's atmosphere.

1 Which is the odd one out in each of the following? Explain your answer.
  a kerosene, petrol, diesel, wood
  b polythene, PVC, aluminium, polystyrene
  c crude oil, olive oil, palm oil, sunflower oil
  d nitrogen, carbon dioxide, hydrogen, oxygen

2 Make a list of all the things used to make a packet of crisps.

3 Suggest what might happen to the packet once the crisps have been eaten.

Cracking hydrocarbons

By the end of this topic you should be able to:

- describe what cracking is
- explain that the products of cracking are used as fuels and to make plastics
- describe what alkenes are and how they can be represented.

A Plastics are made from crude oil.

Crude oil is mainly a mixture of **hydrocarbons**. These are substances which only contain hydrogen and carbon. To be useful the mixture must be separated. This is done by **fractional distillation** at an oil refinery. Each separate part is called a **fraction**. The fractions have many uses, for example as petrol for cars.

All the fractions are in demand but some, such as petrol, are in greater demand than supply. In other words, more petrol is needed than we produce by the fractional distillation of crude oil. Generally, the fractions containing fewer carbon atoms are in short supply.

Other fractions, such as kerosene which is used for fuel in aeroplanes, is in greater supply than demand. This is generally the case for fractions containing more carbon atoms.

To solve the supply and demand problem, longer hydrocarbons can be broken down. This is done by **cracking** them into smaller hydrocarbons. The hydrocarbons are heated to vaporise them (turn them into gases). They are then passed over a hot **catalyst**. A catalyst is a chemical that speeds up a reaction but does not get used up. Cracking is a **thermal decomposition** reaction because molecules are broken down by heating them.

1 What is crude oil made up of?

2 What can be done to crude oil to make it useful?

3 Give three uses for chemicals produced from crude oil.

4 Name one fraction of crude oil which is in greater demand than supply.

5 Name one fraction of crude oil which is in greater supply than demand.

6 **a** What is cracking?
b Why is cracking used?

7 **a** What are the two different types of products we get by cracking crude oil?
b Give a use for each of these products.

The hydrocarbons in crude oil are called **alkanes**. Cracking produces shorter alkanes and **alkenes**. The shorter alkanes are used to help meet the demands we have for fuel. The alkenes are used to make plastics.

```
   H  H  H  H  H  H  H  H  H  H               H  H  H  H  H  H  H  H          H  H
   |  |  |  |  |  |  |  |  |  |               |  |  |  |  |  |  |  |          |  |
H—C—C—C—C—C—C—C—C—C—C—H    →    H—C—C—C—C—C—C—C—C—H    +    C=C
   |  |  |  |  |  |  |  |  |  |               |  |  |  |  |  |  |  |          |  |
   H  H  H  H  H  H  H  H  H  H               H  H  H  H  H  H  H  H          H  H
```

longer alkanes → shorter alkanes (used as fuels) + alkenes (used to make plastics)

B Cracking.

Alkanes are **saturated hydrocarbons**. These contain no C=C double bonds. Alkenes are **unsaturated hydrocarbons** which contain one or more C=C double bonds.

```
 H  H            H  H  H
 |  |            |  |  |
 C==C            C==C——C——H
 |  |            |     |
 H  H            H     H
```

ethene C_2H_4

propene C_3H_6

C Alkenes.

Alkenes have the general formula C_nH_{2n}.

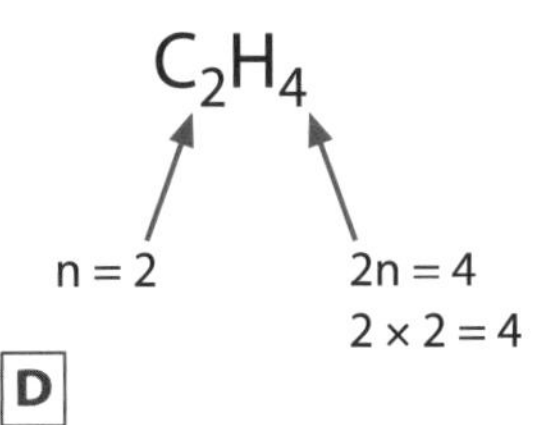

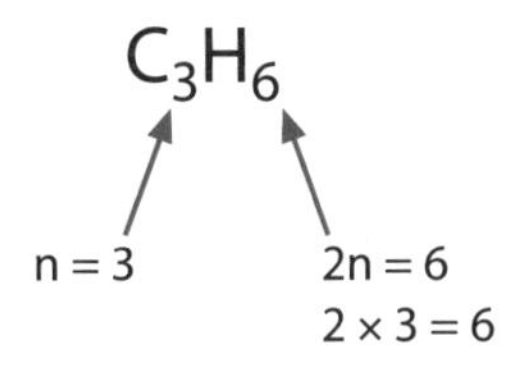

D

A simple test for unsaturated hydrocarbons uses bromine water. Bromine water is a yellow–orange colour. It turns colourless when it reacts with unsaturated hydrocarbons. However, if it is added to a saturated hydrocarbon, there is no reaction and it stays a yellow–orange colour.

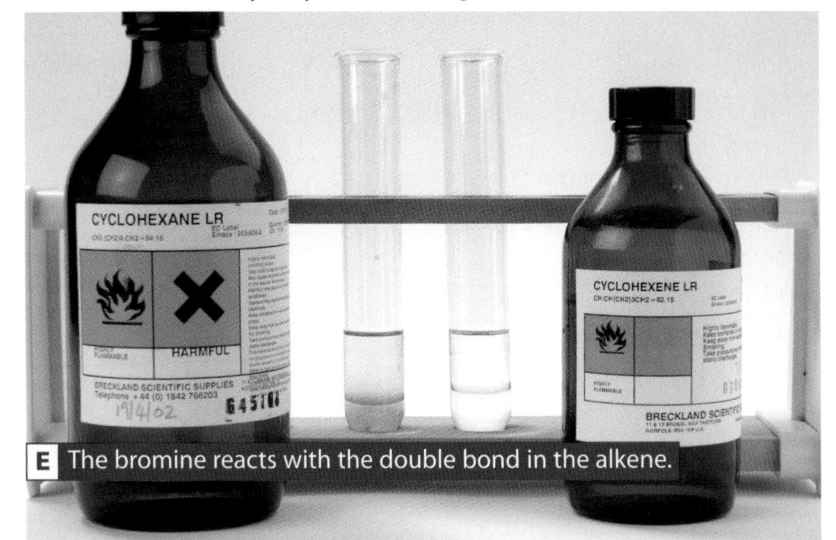

E The bromine reacts with the double bond in the alkene.

8 What is an unsaturated molecule?

9 Butene is an alkene that contains four C atoms.
- **a** Draw the structure of butene.
- **b** Give the formula of butene.

10 If kerosene is cracked, one of the products is ethene. Ethene is an alkene which contains two carbon atoms.
- **a** Explain why kerosene is cracked.
- **b** Describe how kerosene is cracked.
- **c** Give the formula of ethene.
- **d** Describe a test to show that ethene is unsaturated. Give the result of the test in your answer.

Making polymers

By the end of this topic you should be able to:

- describe what a polymer is
- explain how polymers are made from alkenes
- recall that polymers have many different uses.

Polymers (plastics) are used to make many things, including bags, bottles, CDs, DVDs and the casing of electrical items such as computers and mobile phones.

A Some products made from polymers.

Polymers are large molecules made from lots of small molecules called **monomers** joined together. The process in which monomers join together is called **polymerisation**. Many common polymers, such as polythene, are made from alkenes. This is because the double bond in alkenes can be broken and used to join the molecule to another alkene molecule.

When ethene is heated at very high pressures, the ethene molecules react with each other to form the polymer **poly(ethene)**. This polymer is better known as polythene and is used to make things like plastic bags and plastic bottles.

1 What is a polymer?

2 What are monomers?

3 What happens during polymerisation?

4 How does the structure of alkenes allow them to form polymers?

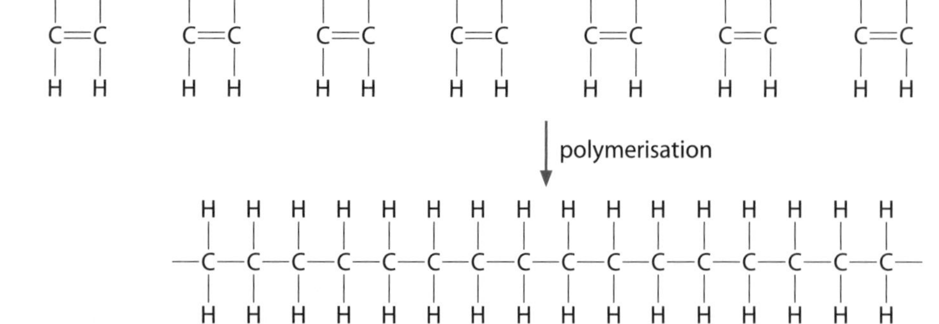

B How poly(ethene) is made.

An equation can be written for the polymerisation of ethene, as shown in diagram C. The number of molecules that join together is very large. It is often several thousand but the exact number varies. We can write the formula as shown in diagram C, where n means a large number.

$$n \begin{array}{c} H \quad H \\ | \quad\ | \\ C{=}C \\ | \quad\ | \\ H \quad H \end{array} \longrightarrow \left[\begin{array}{c} H \quad H \\ | \quad\ | \\ -C-C- \\ | \quad\ | \\ H \quad H \end{array} \right]_n$$

C Equation for the formation of poly(ethene).

5 What is the common name of poly(ethene)?

6 a Which alkene is used to make poly(ethene)?
b What type of reaction is used to make poly(ethene)?

7 Give two uses for poly(ethene).

Poly(propene) is formed from the polymerisation of the alkene propene. Poly(propene) is used to make plastic crates, bins and rope.

$$n \begin{array}{c} CH_3 \ H \\ | \quad\ | \\ C{=}C \\ | \quad\ | \\ H \quad H \end{array} \longrightarrow \left[\begin{array}{c} CH_3 \ H \\ | \quad\ | \\ -C-C- \\ | \quad\ | \\ H \quad H \end{array} \right]_n$$

D Equation for the formation of poly(propene).

8 Give two uses of poly(propene).

9 Draw a diagram to show eight molecules of propene reacting together to form poly(propene).

Many other polymers can be made from polymers containing C=C double bonds, including PVC, Teflon, polystyrene and Perspex.

10 Write an equation for the formation of poly(chloroethene), also called PVC, from chloroethene. The structure of chloroethene is shown in E.

$$\begin{array}{c} C \quad Cl \\ | \quad\ | \\ C{=}C \\ | \quad\ | \\ H \quad H \end{array}$$

E

11 a Describe how polymers can be made from alkenes. Include the following words or phrases: alkene, C=C double bond, join, polymerisation, long chain.
b Draw an example of a polymer being made from alkenes.

C1b.3

Properties of polymers

By the end of this topic you should be able to:

- compare the different properties of polymers
- explain how the properties of a polymer can depend on how it is made.

There are many different polymers in use today. These polymers have different properties and so they are used for different things.

Polymer	Common name	Properties	Uses	
Poly(ethene)	polythene	flexible	bags, cling film	
Poly(propene)	polypropene	flexible	crisp packets	
Poly(chloroethene)	PVC	tough, hard	window frames, gutters, pipes	
Poly(tetrafluoroethene)	Teflon/PTFE	tough, slippery	frying pan coatings, stain-proof carpets	
Poly(methyl 2-methylpropenoate)	Perspex	tough, hard, clear	shatterproof windows	

A Polymers and their properties.

1 Use the table to help you explain why
a polythene is suitable for making bags
b Teflon is used for making frying pans.

2 What is the advantage of making window frames out of PVC rather than wood?

3 In some cricket grounds, the pavilion windows are made of Perspex and not glass. Explain why.

The properties of a polymer depend on what it is made from and the conditions in which it is made. For example, there are two forms of polythene: high-density poly(ethene) (HDPE) and low-density poly(ethene) (LDPE).

Polymer	How it is made	Difference in structure	Properties	Uses
Low-density poly(ethene), LDPE	temperature: 200 °C pressure: 2000 atm catalyst: trace of oxygen	molecules are loosely packed due to molecules being more branched	flexible, soft	bags, cling film
High-density poly(ethene), HDPE	temperature: 60 °C pressure: 2 atm catalyst: Ziegler-Natta	molecules are tightly packed as they are less branched (this makes the polymer more rigid)	stiffer, harder than LDPE	buckets, bottles

B Differences between poly(ethene) polymers.

4 Why are plastic buckets not made out of low-density poly(ethene)?

5 Why is cling film not made out of high-density poly(ethene)?

6 High- and low-density poly(ethene) are both made from ethene. What is done differently to make these two polymers with different properties?

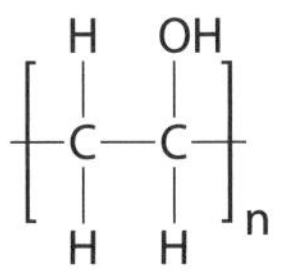

C Poly(ethenol).

Slime is a thick, sticky, slippery substance. One type of slime can be made from the polymer poly(ethenol).

Poly(ethenol) is an unusual polymer because it dissolves in water. One of its uses is to make hospital laundry bags. The bag containing the washing can be put in the washing machine and the bag then dissolves in the water.

Poly(ethenol) reacts with borax to make slime. The borax links the polymer chains together.

Viscosity means how easily a substance flows. A viscous substance, like syrup, flows slowly. In the reaction between poly(ethenol) and borax, the more borax that is used, the more linking there is. The more links there are the more viscous the slime becomes. This is because the poly(ethenol) chains cannot move around so easily.

D A soluble laundry bag used in hospitals.

7 Explain why hospital laundry bags are made from poly(ethenol).

8 Slime is very viscous. What does this mean?

9 When poly(ethenol) reacts with borax, slime is formed. Explain why slime becomes more viscous if more borax is used.

10 Explain why the conditions used to make a polymer are important. Give an example.

P P D P P

Using polymers

By the end of this topic you should be able to:

- describe how new polymers are being developed with new uses.

Many new uses are being found for polymers, and new polymers are being made with unusual properties.

Smart materials are materials which have one or more properties that change in different conditions.

Shape memory polymers are smart materials. They change shape as the temperature changes but can go back to their original shape again. Two uses of shape memory polymers are in heat-shrink wrapping used for packaging and in heat-shrink tubing used to cover bundles of electrical wires.

Other smart materials are polymers that change colour with changes in temperature or light. For example, there are plastic bowls and spoons for baby food which change colour if the food is too hot. There are also light-sensitive lenses for spectacles that darken in bright light.

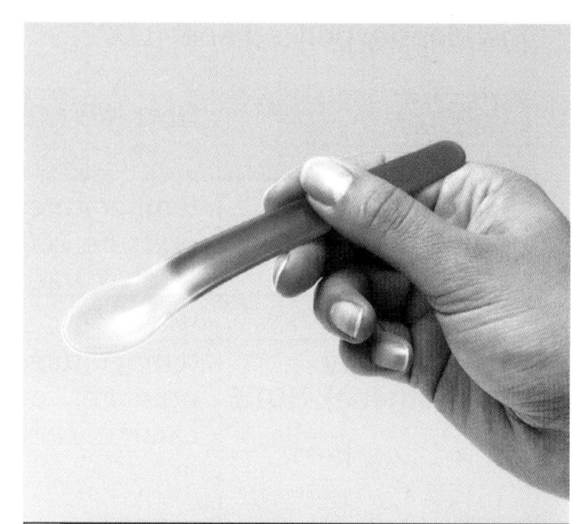

A The colour of the spoon changed from purple to pink to show that the food was too hot.

1 What is a smart material?

2 What is a shape memory polymer?

3 Give four different uses for smart materials.

Hydrogels are polymers that can absorb a lot of water. One use is in nappies where the hydrogel is built into the nappy to absorb urine to prevent nappy rash.

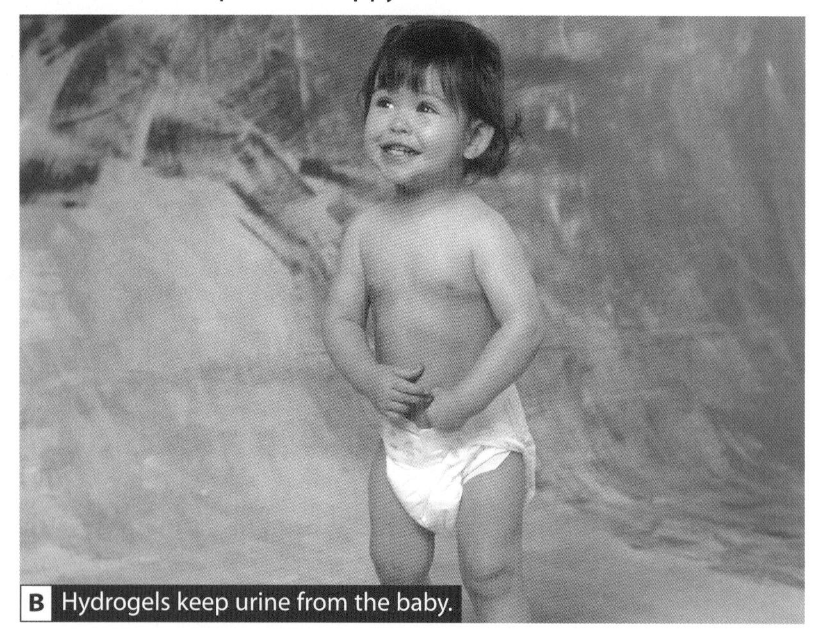

B Hydrogels keep urine from the baby.

4 What is a hydrogel?

5 Why are hydrogels used in nappies?

Light, waterproof coatings for fabrics have been developed that will keep water out but allow moisture from sweat to escape.

For many years, dentists have used metals to fill cavities in teeth. Over the last few years new tooth-coloured materials have been developed using polymers. These are mixtures of chemicals that include a monomer and chemicals to start the polymerisation reaction. The mixture is placed in the cavity and a very bright blue light is shone onto the mixture in the tooth. This starts the polymerisation reaction.

C A breathable, waterproof jacket.

6 Why do people prefer to have polymer fillings in front teeth instead of metal fillings?

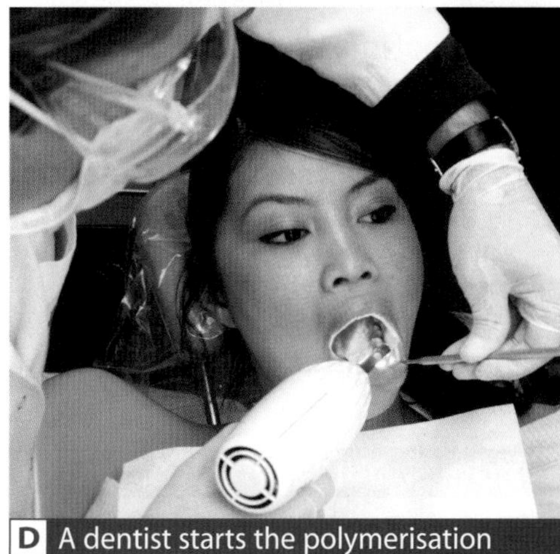

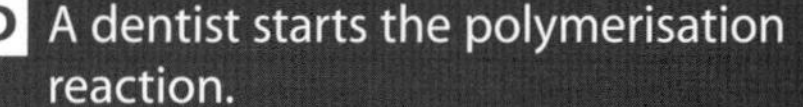

D A dentist starts the polymerisation reaction.

Plastics have been used in many ways in recent years for packaging. Newer developments include food packaging that changes colour if the food starts to go off, the addition of antimicrobial chemicals to prevent bacterial growth and biodegradable plastics that decay naturally over time.

E Different types of food packaging.

7 a Give two recent developments in packaging materials.
b Explain why these are useful developments.

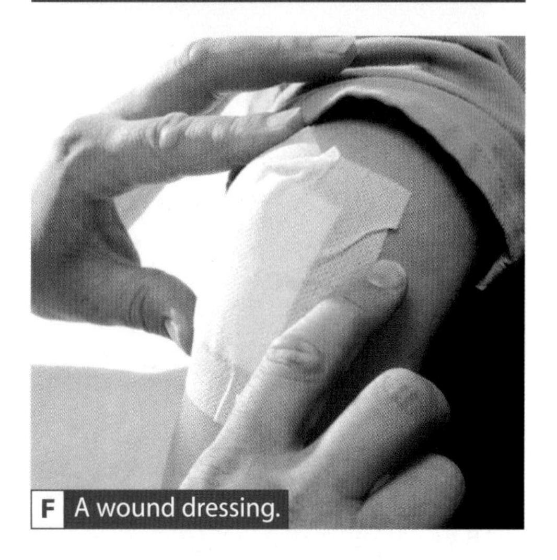

F A wound dressing.

Wound dressings are needed to protect wounds while they heal. New polymers have improved dressings which can now include antibacterial barriers, hydrogels and waterproof but breathable films.

8 What advantage does a waterproof coating on a wound dressing have?

9 What advantage does an antibacterial coating on a wound dressing have?

10 There have been many developments in the production and uses of polymers in recent years. Choose two examples from this topic, and explain why these developments are helpful to our everyday lives.

C1b.5

Disposal of polymers

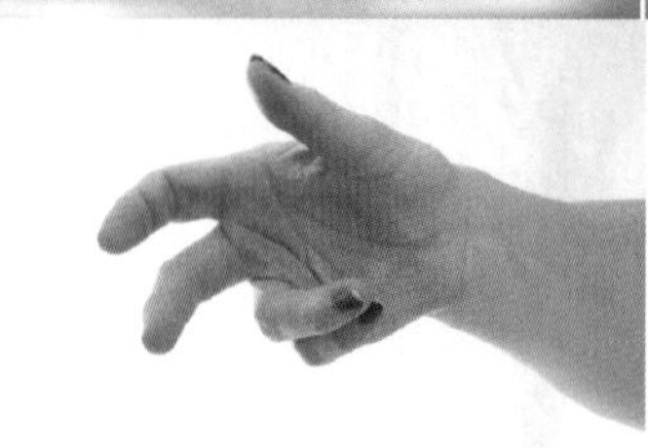

By the end of this topic you should be able to:

- describe the three methods to dispose of polymers
- explain the advantages and disadvantages of each method
- evaluate the advantages and disadvantages of using products from crude oil.

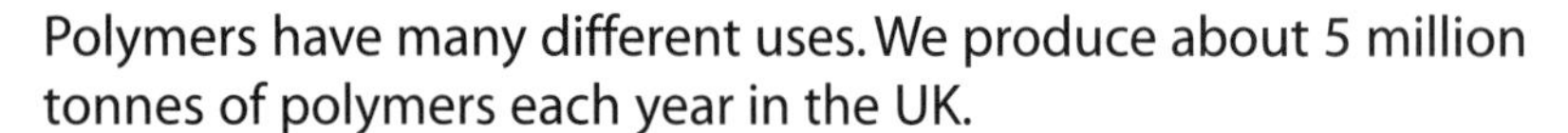

Polymers have many different uses. We produce about 5 million tonnes of polymers each year in the UK.

Most polymers are buried in **landfill** sites. This is easy and cheap, but can cause many problems. It is becoming difficult to find places to put more landfill sites as we bury more and more rubbish. Also, most polymers are not **biodegradable**. This means that microorganisms cannot break them down so they will not **decompose** (rot away) if they are buried.

New polymers are being developed that are broken down by microorganisms (**biodegradable**). However, these are more expensive.

A Throwing away plastic.

How should polymers be disposed of?

B Landfill site.

Polymers burn well and many are burned in **incinerators**. The heat released can also be used to generate electricity. However, carbon dioxide (CO_2) is formed when polymers are burned. This greenhouse gas is thought to be contributing to global warming and other environmental problems. Unless the conditions in the incinerator are carefully controlled, toxic gases including carbon monoxide (CO), hydrogen chloride (HCl) and hydrogen cyanide (HCN) can also be produced.

1 How are most polymers disposed of?

2 a Most polymers are not biodegradable. What does this mean?

b Why is this a problem for burying polymers in landfill?

3 Give two disadvantages of burning polymers.

4 Give two advantages of burning polymers.

C Waste incinerator.

Most polymers can be **recycled**. This means that less crude oil and energy are used in making them and it avoids the problems caused by burying or burning waste ones. Polymers are often recycled by chopping up the polymer into pellets which can be melted and moulded into new products. There are new developments that allow some polymers to be broken back down into monomers. New polymers can then be made from these.

Polymers must be separated out into their different types to be recycled. Most polymer products have a symbol to show which polymer they are made from. However, the sorting has to be done by hand which is time-consuming and expensive. New methods are being developed to sort polymers by machines.

Symbol	1 PET	2 HDPE	3 PVC	4 LDPE	5 PP	6 PS	7 OTHER
Polymer	polyethylene terephthalate	high-density polythene	poly(vinyl chloride)	low-density polythene	poly(propene)	polystyrene	other polymers
Uses	some bottles, food trays, duvet fillings	some bottles, buckets	soft toys, window frames	cling film, bags	crisp packets, carpet, ropes	egg boxes, foam packaging	

D Symbols and uses of polymers.

5 Draw a flow diagram to show what happens to polymers when we have finished with them.

6 a What must be done to polymers before they can be recycled?
b How is this usually done?
c How are advances in technology helping to solve this problem?

7 There are about 60 million people in the UK. 17.5 thousand million supermarket bags are used every year. Calculate the average number of bags used per person in the UK each year.

8 List as many ways as possible that you can reduce the problems caused by using polymers.

9 Draw a table to list the three ways of disposing of polymers. Include columns to show:
a what is done in each method
b the advantages and disadvantages of each method
c whether each method has any social, economic or environmental issues.

C1b.6

Making ethanol

By the end of this topic you should be able to:

- describe how ethanol can be made by fermenting sugars
- describe how ethanol can also be made from ethene and steam
- compare the advantages and disadvantages of the two ways of making ethanol.

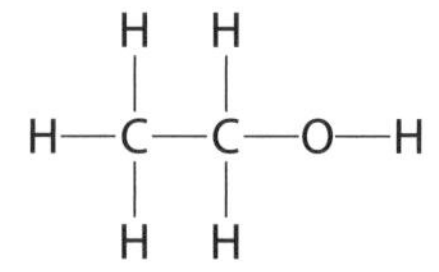

A The structure of ethanol.

There are many different types of alcohol. The commonest one is ethanol. It is found in alcoholic drinks.

Ethanol has many uses besides alcoholic drinks. It is used as a solvent to make many common substances such as detergents and pharmaceutical products (medicines). It can also be used as a fuel. In some countries, such as Brazil, it is used as fuel for cars.

B Ethanol used to fuel cars in Brazil.

1 Give three uses for ethanol.

2 What is the formula of ethanol?

3 Is ethanol a hydrocarbon? Explain your answer.

Over 90% of the world's ethanol is produced from crops such as sugar cane, sugar beet, corn, rice and maize. This is done by fermentation. These are **renewable** raw materials because we can grow more to replace those which have been used. The ethanol made by fermentation is not pure, it also contains water. Fractional distillation is used to make pure ethanol from the mixture.

Alcoholic drinks are made by fermentation. Different drinks are made by fermenting different crops. For example, wine is made from grapes and beer from malted barley. Yeast and water are added to the crop and the mixture left, in the absence of air, for fermentation to produce the drink.

C Alcoholic drinks made by fermentation.

The other main method of producing ethanol is by reaction of ethene with steam. Ethene is made from crude oil. Crude oil is a **non-renewable** resource which means that it cannot be replaced when we use it. However, the ethanol produced this way is pure.

	Fermentation of carbohydrates	Reaction of ethene with steam
Raw materials	source of carbohydrates (e.g. sugar cane, sugar beet, corn, rice and maize)	crude oil
Type of raw materials	renewable	non-renewable
Reaction	sugar → ethanol + carbon dioxide $C_6H_{12}O_6 \rightarrow 2\ C_2H_5OH + 2\ CO_2$	ethene + steam → ethanol $CH_2{=}CH_2 + H_2O \rightarrow C_2H_5OH$
Conditions	30–40 °C microorganisms (often yeast) required (contain **enzymes**) in water anaerobic (no oxygen present)	300 °C high pressure (60–70 atm) concentrated phosphoric acid (catalyst)
Type of process	batch process (stop–start process which is labour intensive, i.e. requires a lot of workers)	continuous process (process kept running 24 hours a day, 7 days a week; i.e. less labour intensive)
Comparison of reaction rates (speed)	slow reaction	fast reaction
Comparison of ethanol made	impure ethanol (purified by fractional distillation)	pure ethanol

D

4 **a** What are the two main ways of producing ethanol?
b Which of these two methods is used most?

5 Which method for producing ethanol:
a is fastest
b uses renewable raw materials
c takes place at lower temperatures and pressures
d gives purer ethanol
e is a continuous process
f has lower labour costs?

6 Explain the difference between a batch process and a continuous process.

7 A local company makes ethanol from ethene.
a Describe how this is done.
b Write a letter to the company to persuade them to switch to making ethanol by fermentation.

C1b.7

Vegetable oils

By the end of this topic you should be able to:

- describe how vegetable oils can be extracted from some fruits, seeds and nuts
- explain that vegetable oils provide energy and nutrients
- compare the advantages and disadvantages of using vegetable oils to produce fuels.

A Vegetable oils are found in fruits, seeds and nuts.

We use large amounts of vegetable fats and oils. Vegetable fats are solids at room temperature whereas vegetable oils are liquids.

1 What is the difference between a vegetable fat and a vegetable oil?

2 Name three sources of vegetable fats and oils.

B Crisps are made using vegetable oil.

Vegetable oils are important foods as they provide energy and contain nutrients such as vitamin E, minerals and essential fatty acids. Some vegetable oils are used to cook food. For example, crisps are made by frying potato slices in vegetable oil. Other foods, such as margarine, salad cream and mayonnaise, contain vegetable fats and oils.

C Foods containing vegetable oil.

3 Why are vegetable oils important in your diet?

4 Give three examples of foods which contain vegetable oils.

The oils in fruits, seeds and nuts have to be extracted to be useful. For some vegetable oils, such as olive oil, this is a simple process which involves pressing (crushing) the seed or fruit. Sometimes there is water as well as oil in the seed or fruit. The water and oil can be easily separated as they form an **emulsion**. The emulsion is spun at high speed which separates the oil and water into two layers. The oil is then purified.

Other vegetable oils, such as sunflower oil, are more difficult to extract. The fruit or seeds are pressed but the oil must be removed by adding a **solvent** to dissolve the oil before it can be purified. The solvent is then removed by **distillation** and reused. Some of the nutrients are lost when vegetable oils are extracted in this way.

Vegetable oils are also used to make non-food products such as detergents and **biodiesel**. Biodiesel can be used in cars, buses and lorries instead of diesel from crude oil. It is mainly made from rapeseed and soybean oils.

5 Draw a flow diagram to show how:
 a olive oil is extracted
 b sunflower oil is extracted.

6 Why does extracted olive oil contain more nutrients than extracted sunflower oil?

Should we use biodiesel as a fuel?

D This diesel contains 5% biodiesel.

Most diesel engines can run on biodiesel. It has the advantage of being made from renewable raw materials and is a cleaner fuel than diesel. It is **carbon neutral**, which means that it releases the same amount of carbon dioxide into the atmosphere when it is burned as the crops took in when they grew. However, it is much more expensive to produce than diesel made from crude oil. Its production also carries the risk of fire and explosion so has to be done carefully.

7 What is biodiesel?

8 Give three advantages of using biodiesel rather than diesel made from crude oil.

9 Give two disadvantages of using biodiesel rather than diesel made from crude oil.

10 Summarise in a paragraph:
 a the use of vegetable oils in food
 b the extraction of vegetable oils
 c the use of vegetable oils as fuels.

Emulsions

By the end of this topic you should be able to:

- recall that oils do not dissolve in water
- describe how emulsions can be made by mixing oils and water
- recall that emulsions are thicker than oil or water
- describe the many uses of emulsions that depend on their special properties.

Vegetable oils do not dissolve in water. If they are mixed together they separate and form two layers. However, if an **emulsifier** is added they will mix and form an **emulsion**. Milk, butter, ice cream, mayonnaise, chocolate and hand cream are all examples of emulsions.

In an emulsion, droplets of one liquid are spread throughout the other liquid. Some emulsions such as milk are oil-in-water emulsions. This means that tiny oil droplets are spread throughout water. Other emulsions, such as margarine are water-in-oil emulsions where water droplets are spread throughout the oil.

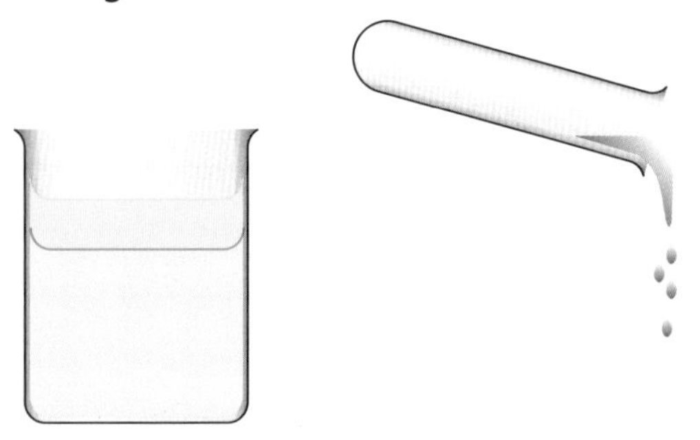

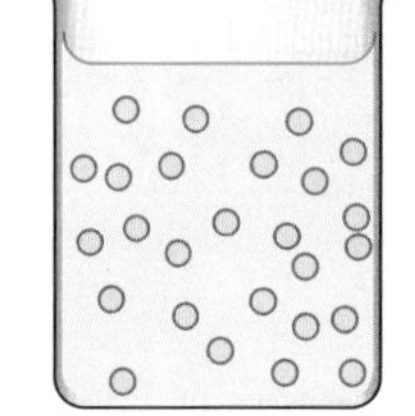

Oil and water form separate layers.

Emulsifier is added.

Emulsifier molecules form a layer on surface of oil droplets allowing them to mix with the water.

C Oil-in-water emulsion.

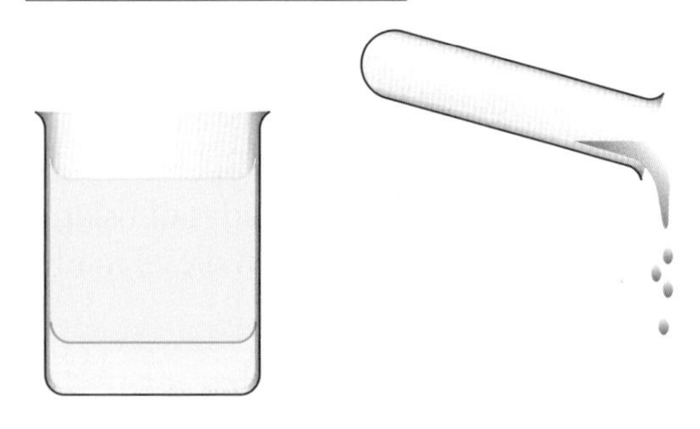

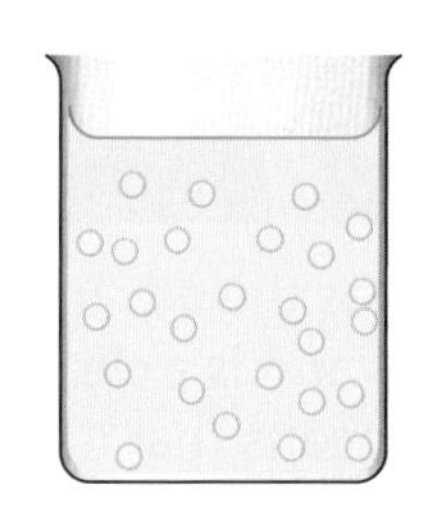

Oil and water form separate layers.

Emulsifier is added.

Emulsifier molecules form a layer on surface of water droplets allowing them to mix with the oil.

D Water-in-oil emulsion.

A Mayonnaise with emulsifier.

B Mayonnaise without emulsifier.

1. Are vegetable oils soluble in water?
2. What happens if an oil and water are mixed together?
3. What is an emulsion?
4. What can be added to a mixture of an oil and water to allow them to mix?
5. Explain how an emulsifier allows a mixture of water and vegetable oil to form a water-in-oil emulsion.

Emulsions are thicker than water or oil. This means that they provide a better texture and appearance and can be used as coatings. For example, neither water nor oil can be spread on bread because they would soak into it. However, butter and margarine, which are emulsions, can be spread and form a coating on the bread.

Some emulsions occur naturally, such as milk. Milk contains water and milk fat. Tiny droplets of milk fat are spread throughout the water. Protein molecules act as the emulsifier. In full-fat milk this emulsion is not completely stable because the fat separates and rises to the top as cream. If it is shaken the emulsion reforms. Other natural emulsions include egg yolks, where droplets of egg oil are spread throughout water, with albumen and lecithin acting as emulsifiers.

Some emulsions are manufactured. Mayonnaise is made by mixing vegetable oil, vinegar, egg yolks and seasoning. The albumen and lecithin in the egg yolks act as emulsifiers. Ice cream is a frozen emulsion.

6 Give four ways in which emulsions can be more useful than an oil or water alone.

	Oil-in-water emulsion	**Water-in-oil emulsion**
Natural emulsion	milk, egg yolk	
Man-made emulsion	mayonnaise, ice cream, salad cream	butter, margarine, moisturiser, skin cream

E Examples of emulsions.

7 What is the emulsifier in milk?

8 Name a manufactured water-in-oil emulsion.

9 Name a manufactured oil-in-water emulsion.

10 Margarine is an example of an emulsion. Use it to explain:
- **a** what an emulsion is
- **b** how it forms
- **c** the properties that emulsions have that oil and water do not.

Hydrogenation of vegetable oils

By the end of this topic you should be able to:

- explain the difference between saturated and unsaturated fats and oils
- describe how unsaturated fats can be detected
- describe how vegetable oils can be hardened using hydrogen
- describe why hydrogenated vegetable oils are useful.

A Common vegetable oils.

There are many different fats and oils, including sunflower oil, olive oil, cod liver oil and tallow (beef fat). Many foods contain a large amount of fats or oils. Vegetable oils are oils which come from plants. These include sunflower oil and olive oil.

Fats and oils have a similar structure. They are made up of three fatty acid molecules and one glycerol molecule. These fats and oils are broken down into fatty acids and glycerol during digestion.

Fats and oils are often described as being saturated or unsaturated. **Saturated** molecules do not contain any C=C double bonds. **Unsaturated** molecules contain one or more C=C double bonds. The structures of some fats and oils are shown in table B.

Type of fat / oil	Example molecule
Saturated (no C=C double bonds in fatty acids)	$CH_3-CH_2-CH_2-CH_2-CH_2-CH_2-CH_2-CH_2-CH_2-CH_2-CH_2-CH_2-CH_2-CH_2-CH_2-CH_2-CH_2-\overset{O}{\overset{\parallel}{C}}-O-CH_2$ $CH_3-CH_2-CH_2-CH_2-CH_2-CH_2-CH_2-CH_2-CH_2-CH_2-CH_2-CH_2-CH_2-CH_2-CH_2-CH_2-CH_2-\overset{O}{\overset{\parallel}{C}}-O-CH$ $CH_3-CH_2-CH_2-CH_2-CH_2-CH_2-CH_2-CH_2-CH_2-CH_2-CH_2-CH_2-CH_2-CH_2-CH_2-CH_2-CH_2-\overset{O}{\overset{\parallel}{C}}-O-CH_2$
Monounsaturated (one C=C double bond in each fatty acid)	$CH_3-CH_2-CH_2-CH_2-CH_2-CH_2-CH_2-CH_2-CH=CH-CH_2-CH_2-CH_2-CH_2-CH_2-CH_2-CH_2-\overset{O}{\overset{\parallel}{C}}-O-CH_2$ $CH_3-CH_2-CH_2-CH_2-CH_2-CH_2-CH_2-CH_2-CH=CH-CH_2-CH_2-CH_2-CH_2-CH_2-CH_2-CH_2-\overset{O}{\overset{\parallel}{C}}-O-CH$ $CH_3-CH_2-CH_2-CH_2-CH_2-CH_2-CH_2-CH_2-CH=CH-CH_2-CH_2-CH_2-CH_2-CH_2-CH_2-CH_2-\overset{O}{\overset{\parallel}{C}}-O-CH_2$
Polyunsaturated (more than one C=C double bond in each fatty acid)	$CH_3-CH_2-CH_2-CH_2-CH_2-CH=CH-CH_2-CH=CH-CH_2-CH_2-CH_2-CH_2-CH_2-CH_2-CH_2-\overset{O}{\overset{\parallel}{C}}-O-CH_2$ $CH_3-CH_2-CH=CH-CH_2-CH=CH-CH_2-CH=CH-CH_2-CH_2-CH_2-CH_2-CH_2-CH_2-CH_2-\overset{O}{\overset{\parallel}{C}}-O-CH$ $CH_3-CH_2-CH_2-CH_2-CH_2-CH=CH-CH_2-CH=CH-CH_2-CH_2-CH_2-CH_2-CH_2-CH_2-CH_2-\overset{O}{\overset{\parallel}{C}}-O-CH_2$

B Differences between saturated and unsaturated fats and oils.

1 What do we mean by the following terms when we talk about fats and oils?
a saturated
b monounsaturated
c polyunsaturated

Fats and oils that are saturated are solids at room temperature. Fats and oils that are unsaturated have lower melting points, so many are liquids at room temperature. The more C=C double bonds they have, the more unsaturated they are, and the lower their melting point is.

2 In what state are the following fats and oils at room temperature?
a saturated
b polyunsaturated

Vegetable oils, which are unsaturated, can be hardened to produce solid fats. Solid fats are useful as spreads for bread (margarine) and for making cakes and pastries. The hardening is done by reacting unsaturated vegetable oils with hydrogen to make saturated or less unsaturated fats. We say they are **hydrogenated.** This is done at about 60 °C in the presence of a nickel catalyst. The hydrogenated oils have higher melting points which is why they are solids at room temperature.

$$CH_3{-}CH_2{-}CH_2{-}CH_2{-}CH_2{-}CH{=}CH{-}CH_2{-}CH{=}CH{-}CH_2{-}CH_2{-}CH_2{-}CH_2{-}CH_2{-}CH_2{-}CH_2{-}\overset{\overset{\displaystyle O}{\|}}{C}{-}O{-}CH_2$$

$$CH_3{-}CH_2{-}CH{=}CH{-}CH_2{-}CH{=}CH{-}CH_2{-}CH{=}CH{-}CH_2{-}CH_2{-}CH_2{-}CH_2{-}CH_2{-}CH_2{-}CH_2{-}\overset{\overset{\displaystyle O}{\|}}{C}{-}O{-}CH \quad \xrightarrow{\text{hydrogen}}$$

$$CH_3{-}CH_2{-}CH_2{-}CH_2{-}CH_2{-}CH{=}CH{-}CH_2{-}CH{=}CH{-}CH_2{-}CH_2{-}CH_2{-}CH_2{-}CH_2{-}CH_2{-}CH_2{-}\overset{\overset{\displaystyle O}{\|}}{C}{-}O{-}CH_2$$

polyunsaturated (liquid at room temperature)

$$CH_3{-}CH_2{-}CH_2{-}CH_2{-}CH_2{-}CH_2{-}CH_2{-}CH_2{-}CH_2{-}CH_2{-}CH_2{-}CH_2{-}CH_2{-}CH_2{-}CH_2{-}CH_2{-}CH_2{-}\overset{\overset{\displaystyle O}{\|}}{C}{-}O{-}CH_2$$

$$CH_3{-}CH_2{-}CH_2{-}CH_2{-}CH_2{-}CH_2{-}CH_2{-}CH_2{-}CH_2{-}CH_2{-}CH_2{-}CH_2{-}CH_2{-}CH_2{-}CH_2{-}CH_2{-}CH_2{-}\overset{\overset{\displaystyle O}{\|}}{C}{-}O{-}CH$$

$$CH_3{-}CH_2{-}CH_2{-}CH_2{-}CH_2{-}CH_2{-}CH_2{-}CH_2{-}CH_2{-}CH_2{-}CH_2{-}CH_2{-}CH_2{-}CH_2{-}CH_2{-}CH_2{-}CH_2{-}\overset{\overset{\displaystyle O}{\|}}{C}{-}O{-}CH_2$$

saturated (solid at room temperature)

C Using hydrogenation to convert oils to fats.

3 Why are polyunsaturated vegetable oils not suitable for using as a spread on bread?

4 How are polyunsaturated vegetable oils hardened?

5 Give three uses of the hardened vegetable oils.

6 Write a paragraph to explain how and why some vegetable oils are hardened.

P

P

D

P

P

Vegetable oils in our diet

By the end of this topic you should be able to:

- explain the effect of using vegetable oils in foods on diet and health.

A Vegetable oils are used to make crisps.

You need to eat fats and oils. They are a good source of energy and provide nutrients such as essential fatty acids. Eating food containing too many fats and oils is not good for you. It can make you overweight or obese.

Eating too much fatty food is also linked to heart disease. Saturated fats tend to be unhealthier than unsaturated ones. They increase the amount of LDL ('bad') cholesterol in your body which contributes to heart disease. Unsaturated fats, especially polyunsaturated ones, contain more of the essential fatty acids that are good for health. They also contain more HDL ('good') cholesterol which can reduce the risk of heart disease.

Hydrogenation of vegetable oils can produce some unsaturated fats called **trans fats**. These trans fats hardly occur in nature. They are thought to cause more damage to the heart than saturated fats. They are found in many foods and there is pressure on manufacturers to remove them from their products. Many processed and fast foods currently contain a lot of trans fats. As more people become concerned, fewer are being used.

1 Give two reasons why you need to eat fats and oils.

2 Why should you not eat too many fats and oils?

3 Why should you eat fewer saturated fats?

4 Which type of fats are the best for you?

5 How are trans fats produced?

6 What type of foods currently contains a lot of trans fats?

At the moment, most experts agree that we should:

- ☐ Reduce the total amount of fat in our diet.
- ☐ Reduce the amount of saturated fat in our diet.
- ☐ Reduce the amount of trans fats in our diet (produced by hydrogenation of polyunsaturated fats).
- ☐ Raise the proportion of polyunsaturated fats, which are found in vegetable oils, in our diet.

B

Butter and hard margarines contain a lot of saturated fats. Soft margarines contain fewer saturated fats and more polyunsaturated ones. There are also some low-fat spreads that contain more water and less fat (as low as 25% fat).

We can use bromine water to see if a fat or oil is saturated or unsaturated. Bromine water reacts with unsaturated molecules. It is yellow–orange but turns colourless with unsaturated fats. If bromine water is added to a saturated fat it stays yellow–orange.

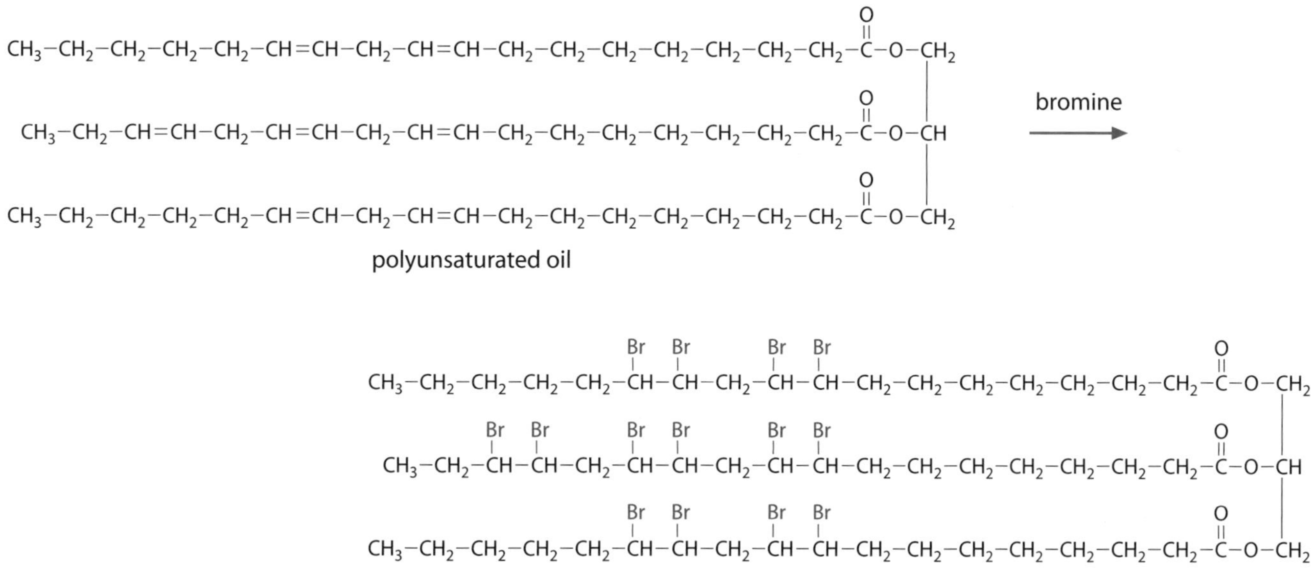

$$CH_3-CH_2-CH_2-CH_2-CH_2-CH=CH-CH_2-CH=CH-CH_2-CH_2-CH_2-CH_2-CH_2-CH_2-CH_2-\overset{O}{\overset{\|}{C}}-O-CH_2$$
$$CH_3-CH_2-CH=CH-CH_2-CH=CH-CH_2-CH=CH-CH_2-CH_2-CH_2-CH_2-CH_2-CH_2-CH_2-\overset{O}{\overset{\|}{C}}-O-CH$$
$$CH_3-CH_2-CH_2-CH_2-CH_2-CH=CH-CH_2-CH=CH-CH_2-CH_2-CH_2-CH_2-CH_2-CH_2-CH_2-\overset{O}{\overset{\|}{C}}-O-CH_2$$

polyunsaturated oil

$$\xrightarrow{\text{bromine}}$$

$$CH_3-CH_2-CH_2-CH_2-CH_2-\overset{Br}{\overset{|}{CH}}-\overset{Br}{\overset{|}{CH}}-CH_2-\overset{Br}{\overset{|}{CH}}-\overset{Br}{\overset{|}{CH}}-CH_2-CH_2-CH_2-CH_2-CH_2-CH_2-CH_2-\overset{O}{\overset{\|}{C}}-O-CH_2$$
$$CH_3-CH_2-\overset{Br}{\overset{|}{CH}}-\overset{Br}{\overset{|}{CH}}-CH_2-\overset{Br}{\overset{|}{CH}}-\overset{Br}{\overset{|}{CH}}-CH_2-\overset{Br}{\overset{|}{CH}}-\overset{Br}{\overset{|}{CH}}-CH_2-CH_2-CH_2-CH_2-CH_2-CH_2-CH_2-\overset{O}{\overset{\|}{C}}-O-CH$$
$$CH_3-CH_2-CH_2-CH_2-CH_2-\overset{Br}{\overset{|}{CH}}-\overset{Br}{\overset{|}{CH}}-CH_2-\overset{Br}{\overset{|}{CH}}-\overset{Br}{\overset{|}{CH}}-CH_2-CH_2-CH_2-CH_2-CH_2-CH_2-CH_2-\overset{O}{\overset{\|}{C}}-O-CH_2$$

C Reaction of an unsaturated oil with bromine.

Fats and oils can be analysed to measure the amount of unsaturation. The more unsaturated a molecule, the more bromine water it reacts with.

7 What do you see when bromine water reacts with unsaturated molecules?

8 Why do polyunsaturated molecules react with more bromine water than monounsaturated molecules?

9 Write a short newspaper article (no more than 200 words long) to explain to the readers:
a the need for some fats and oils in our diet
b the health benefits and problems of different types of fats and oils.

Food additives

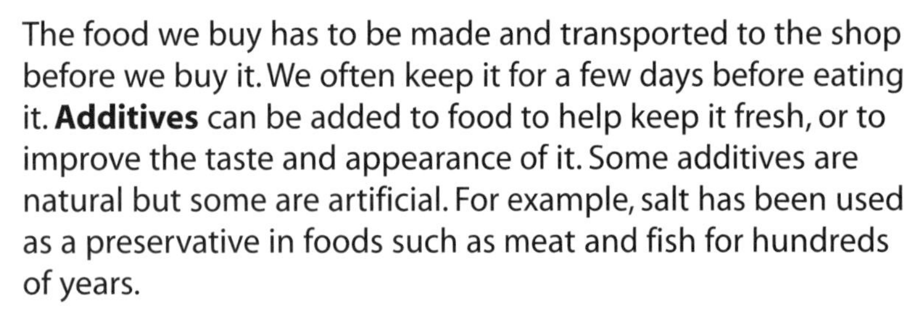

HSW ***By the end of this topic you should be able to:***

- explain why processed foods may contain additives
- recall that additives must be listed in the ingredients and that some have E-numbers
- compare the advantages and disadvantages of using additives in food.

A Ingredients label on a bag of crisps.

The food we buy has to be made and transported to the shop before we buy it. We often keep it for a few days before eating it. **Additives** can be added to food to help keep it fresh, or to improve the taste and appearance of it. Some additives are natural but some are artificial. For example, salt has been used as a preservative in foods such as meat and fish for hundreds of years.

Additives have been given **E-numbers** by the European Food Safety Authority. For an additive to be given an E-number it must first pass safety checks.

Additives	What they do	Examples
Preservatives	Prevent the growth of microbes and stop us getting food poisoning.	E210 benzoic acid (found in fruits) E234 nisin (found in milk and cheese)
Anti-oxidants	Food reacts with oxygen in the air and goes off. Anti-oxidants slow down the reaction with oxygen making the food last longer.	E300 vitamin C E306 vitamin E
Emulsifiers	Allow water and oils to mix in foods such as salad dressings, ice cream and margarine.	E322 lecithin (from egg yolks) E471 mono- or di-glycerides of fatty acids
Colours	Colourings are sometimes added to food to improve the appearance.	E150a caramel (brown colour from heating sugar) E160a beta carotene (orange colour from carrots)
Flavourings	Flavourings are sometimes added to food to improve the taste.	Most flavourings do not have E-numbers because they are controlled by different laws.
Sweeteners	Make the food taste sweeter. Some have very low energy content and are used in slimming (low-calorie) foods.	E951 aspartame E954 saccharin
Stabilisers and thickeners	Help foods to keep their appearance or shape or to make them thicker.	E410 locust bean gum E440 pectin
Acids	Acids are sometimes added to improve flavour.	E330 citric acid E338 phosphoric acid
Anti-caking agents	Added to powders, such as sugar, salt and (powdered) milk to make them flow freely and not go lumpy.	E533b talc E551 silicon dioxide

B

1 What are food additives?

2 Why are additives put in some food?

3 Why are additives given E-numbers?

4 What happens to an additive before it is given an E-number?

There is a lot of debate about food additives. Some studies suggest that some food additives can cause allergies. For example, the preservative sodium sulfite (E221) can trigger asthma attacks. There are very few other examples where the evidence is clear.

Some studies claim that there is a link between some additives and hyperactivity in children. The yellow colouring tartrazine (E102) is a common example but there is not enough evidence to prove a link at the moment. Tartrazine is rarely used now as a food additive in case there is a link.

C E221 can trigger asthma attacks.

D Some food additives might affect behaviour.

In recent years, many food manufacturers have started to put fewer additives in foods. However, many additives are essential and so food manufacturers continue to use them.

5 Why do some people worry about food additives?

6 What harmful effect has E221 (sodium sulfite) been shown to have?

7 What harmful effect does E102 (tartrazine) possibly have?

8 Should we use food additives, restrict them or stop using them altogether? Make a table to show advantages and disadvantages of using them.

Analysis of food additives

By the end of this topic you should be able to:

- explain how chemical analysis can be used to detect food additives
- describe how chromatography can be used to identify colourings in foods.

Scientists need to be able to detect additives in food. Analytical chemists working for the Food Standards Agency (FSA) analyse food to check what is in it. This is to check that the food is safe and that the label identifying the ingredients is correct.

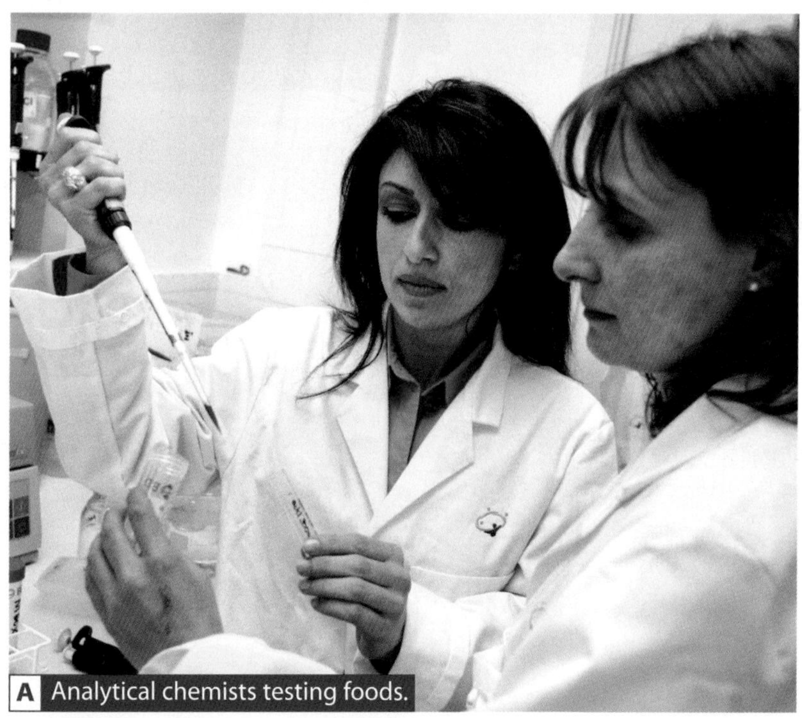

A Analytical chemists testing foods.

1 Which official group analyses food in the UK?

2 What two things do analysts check for when they analyse additives in food?

There are many techniques used by the analysts. One of them is **chromatography**. This can be used to test for colourings in food. Colourings are sometimes added to food to improve the appearance.

B Food labels show the colourings.

Chromatography (see diagram C) separates the colourings out. A sample of the colour from the food is placed on a piece of chromatography paper and then a solvent is added at the bottom of the beaker. The solvent soaks up the paper taking the colourings with it. The more soluble a colouring is, the further up the paper it travels. Different colourings will move different distances up the paper.

3 What can chromatography be used for in food analysis?

4 Explain how chromatography separates the colourings from a food.

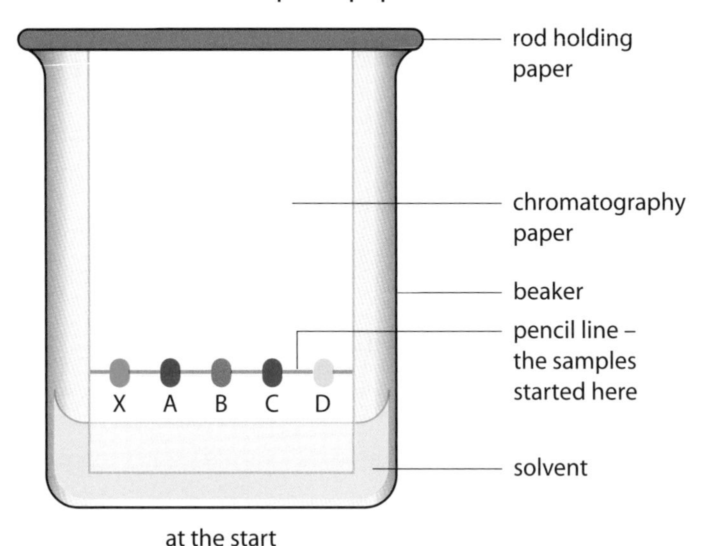

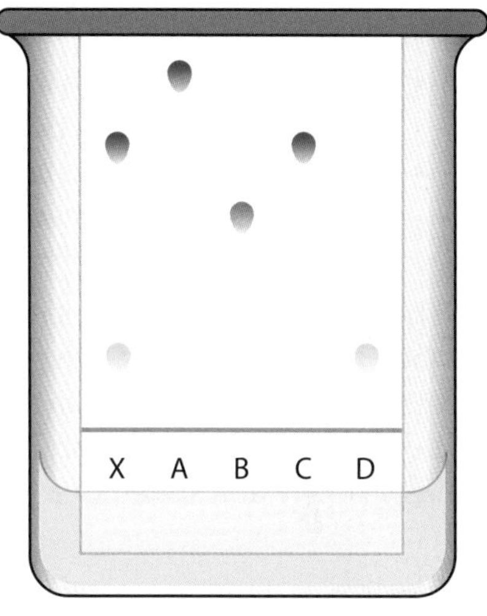

C Chromatography.

The analysis of the colours in some sweets is shown in diagram D. The chromatogram on the left shows some common food colourings. The chromatogram on the right shows the results for the sweets. From this you can tell which food colourings are in which sweets. Some sweets only contain one colouring while others contain a mixture of colourings. For example, the violet sweets contain E120 and E133.

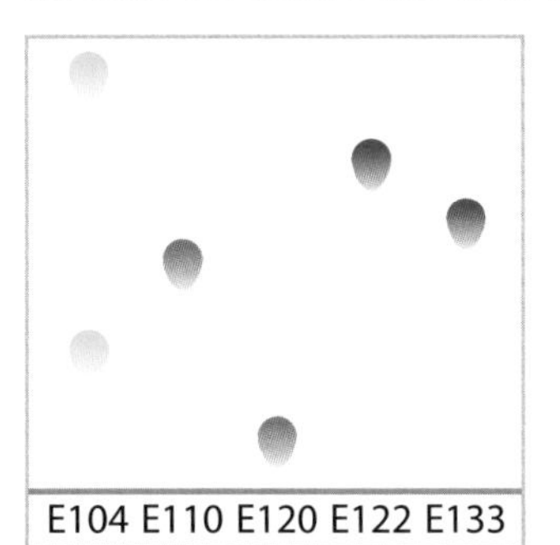

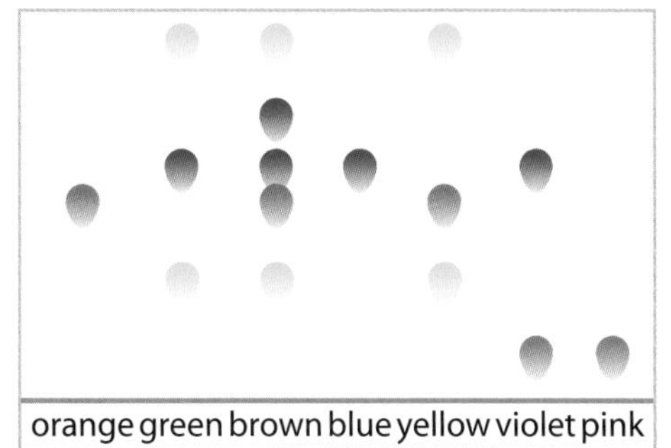

D Analysis of colourings in sweets.

5 Look at the chromatogram of the sweets in D.
- **a** Which sweets contain one colouring?
- **b** **(i)** Which sweet contains the most colourings?
  - **(ii)** Which colour sweet contains the most colourings?
- **c** Which colourings are in the yellow sweets?
- **d** Which colourings are in the green sweets?

6 **a** Describe an experiment you could do to show that a food contains the colouring tartrazine (E102).
- **b** What would you need to test apart from the food?
- **c** How could you tell from the results that the food contains tartrazine?

Tectonic plates

By the end of this topic you should be able to:

- recall that the Earth's crust and upper part of the mantle are cracked into large pieces called tectonic plates
- explain how convection currents within the mantle cause these plates to move slowly
- explain why earthquakes and volcanoes occur at boundaries between plates
- explain why scientists cannot accurately predict when earthquakes and volcanic eruptions will occur.

The Earth is thought to be made up of layers. The outer part of the Earth is called the **lithosphere** and includes the **crust** and the upper part of the **mantle**. It is cracked into a number of huge pieces called **tectonic plates**. These plates are moving very slowly at a speed of a few centimetres each year.

1 What are tectonic plates?

2 How much do tectonic plates move?

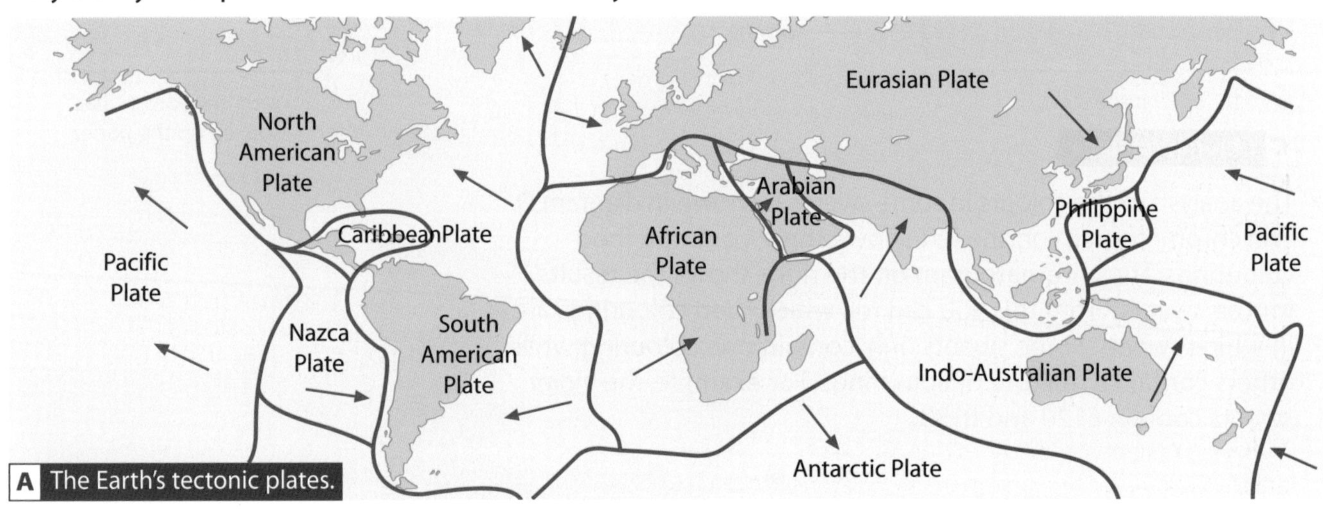

A The Earth's tectonic plates.

The slow movement of tectonic plates is caused by very powerful **convection currents** in the mantle.

The currents are caused by heat released inside the Earth from the natural breakdown (decay) of radioactive atoms.
As rock in the mantle gets hotter it becomes less dense and rises.
It is pushed to the side by more rock where it cools.
The cooling means that it becomes more dense and sinks back down.
It is the sideways movement of the rock that moves the plate above it.

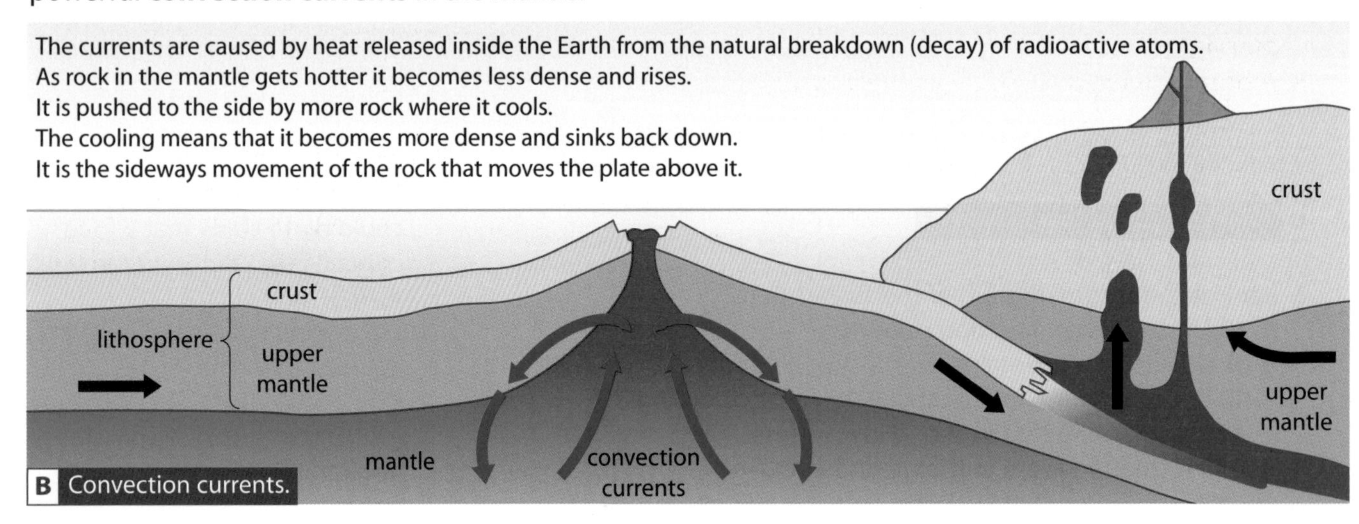

B Convection currents.

Volcanoes and **earthquakes** occur at the boundaries between tectonic plates. As plates move past, over, under or apart from each other, hot magma from the mantle can escape, resulting in a volcano. Friction between the moving plates may stop the plates moving smoothly. If this happens, the plates move in sudden jerks resulting in earthquakes.

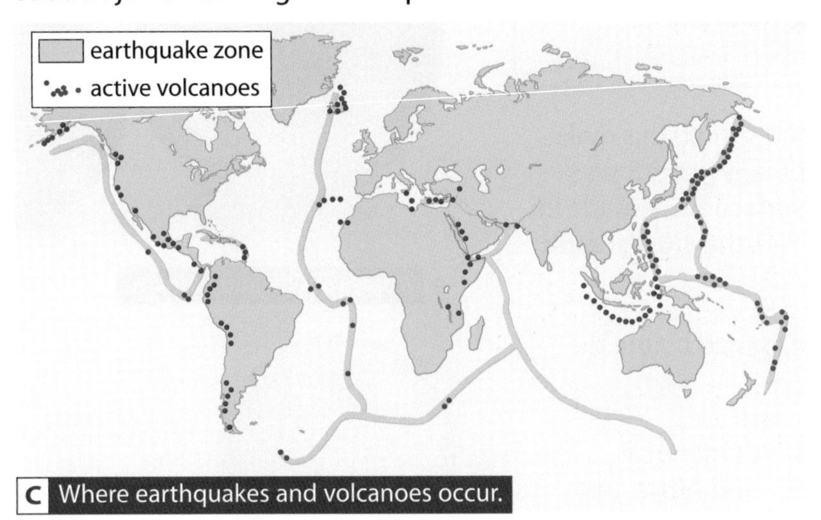

C Where earthquakes and volcanoes occur.

Predicting earthquakes and volcanic eruptions

Many people live in areas at risk from earthquakes. Scientists know where earthquakes are likely and so many buildings in danger zones are built with special foundations to help them to withstand earthquakes.

There are some warning signs before an earthquake:

- small earthquakes (pre-shocks)
- water levels in wells fall
- some animals act strangely.

Volcanic eruptions are much easier to predict than earthquakes. Warning signs before an eruption include:

- increasing temperature of the volcano due to magma moving underground
- rising ground level around the volcano due to the build up of magma
- more sulfur dioxide (SO_2) gas is given out.

When these warning signs appear, people can be moved to safety. However, scientists cannot reliably predict exactly *when* major earthquakes or volcanic eruptions are going to happen. It is difficult to predict when there will be enough pressure for plates to slide past each other or for magma to burst up.

3 Explain why tectonic plates move.

4 What causes convection currents inside the Earth?

P

5 Why do volcanoes and earthquakes develop at plate boundaries?

P

6 Why is it important to be able to predict where and when earthquakes and volcanic eruptions will happen?

7 Which is easier to predict: an earthquake or a volcanic eruption? Explain your answer.

8 Why can scientists not reliably predict exactly when earthquakes or volcanic eruptions will occur?

P

9 Imagine that you are a journalist living in an earthquake zone near a volcano. Write a short newspaper article (no longer than 200 words) explaining:

a why earthquakes and volcanoes occur where you live

b what scientists can do to predict earthquakes and volcanic eruptions.

D

P

P

The theory of continental drift

HSW By the end of this topic you should be able to:

- explain what the theory of continental drift is
- explain why it took many years for the theory of continental drift to be accepted.

As apples get older they start to dry up and shrink. This makes the skin too big and so it wrinkles up. For many years it was believed that the features on the Earth's surface were made in a similar way. As the young, hot Earth cooled, the crust shrank and wrinkled, forming mountains.

1 How did people believe mountains were formed before the theory of plate tectonics?

In 1911 Alfred Wegener read that the fossils of identical creatures had been found in South America and Africa, and that these creatures could not swim. People at the time said that there must have been a piece of land between the continents which was now covered by the Atlantic Ocean.

Wegener came up with the idea of **continental drift**. He believed that the continents were moving around and were once joined together in a big land mass. In 1915, Wegener published a book about his theory.

The main evidence in Wegener's book was:

- the continents appear to fit together like a jigsaw
- the west coast of Africa and the east coast of south America have the same patterns of rock layers.
- these two coasts have the same types of plant and animal fossils; some of these animals are only found in those parts of the world and their fossils show they could not swim
- there could not have been a piece of land connecting South America and Africa, which means they must have been joined together.

A Apples shrink and wrinkle.

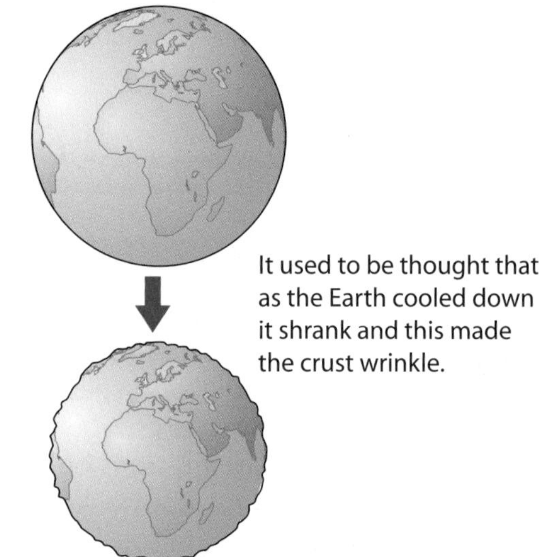

B The shrinking Earth theory.

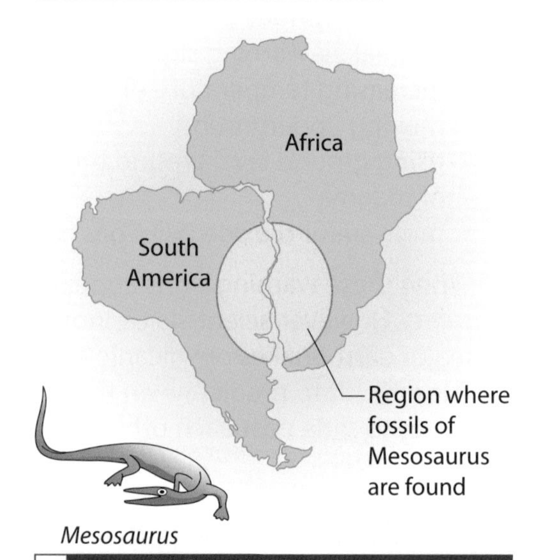

C Position of the continents millions of years ago and the region in which fossils of *Mesosaurus* have been found.

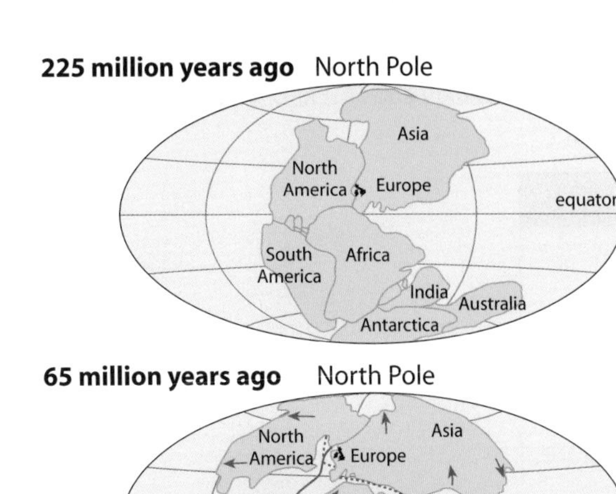

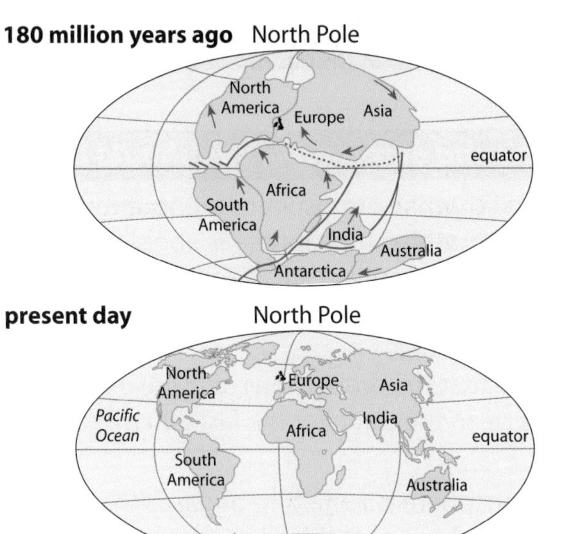

65 million years ago North Pole

North America
Asia
Europe
equator
South America
Africa
India
Australia
Antarctica

present day North Pole

North America
Europe
Asia
Pacific Ocean
India
Africa
equator
South America
Australia
Antarctica

D The moving continents.

When Wegener died in 1930 his ideas had still not been accepted. This was because there was no explanation for how the continents moved and because Wegener was not a geologist.

His ideas were not accepted until the 1960s when the Atlantic Ocean floor was surveyed in detail and the mid-Atlantic ridge was found. This is a range of underwater mountains and volcanoes in the middle of the ocean.

Soon afterwards it was discovered that the rock in the ocean floor is younger than the rock in the continents. The rock closest to the ridge is youngest. It was also found that the magnetic alignment of rocks containing iron-rich minerals was symmetrical either side of the ridge. This new evidence fitted in with the theory of continental drift.

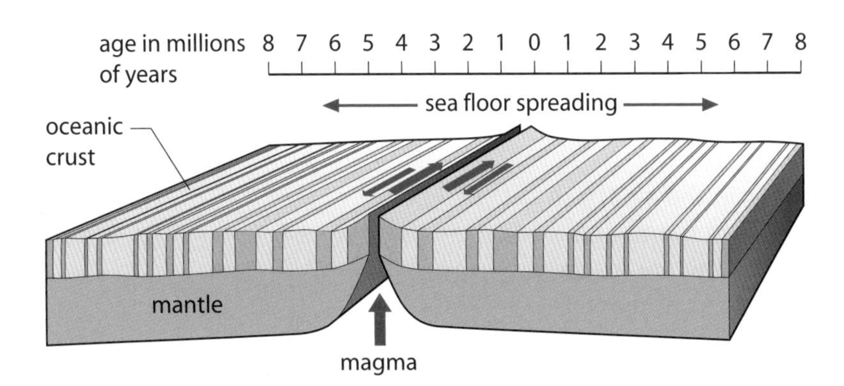

iron rich minerals aligned towards current North pole

iron rich minerals aligned towards current South pole

E Sea floor spreading. This is caused by the Earth's magnetic field reversing its direction every few hundred thousand years.

Scientists also discovered that this movement could be caused by very powerful convection currents in the mantle. This brought about the new theory of plate tectonics which is very similar to Wegener's theory of continental drift.

2 What is meant by continental drift?

3 How do rock patterns support Wegener's ideas?

4 How do fossils of *Mesosaurus* support Wegener's ideas?

5 What other evidence did Wegener use to support his ideas?

6 Give two reasons why Wegener's ideas were not accepted at the time.

7 What evidence was discovered that led to Wegener's ideas being accepted?

8 Imagine you were a scientist living in the 1960s. List all the evidence you have to support the theory of plate tectonics.

The atmosphere

By the end of this topic you should be able to:

- recall that the atmosphere is about four-fifths nitrogen
- recall that the atmosphere is about one-fifth oxygen
- recall that there are small amounts of other gases in the atmosphere including noble gases, carbon dioxide and water vapour
- recall that the proportions of these gases in the atmosphere has been the same for the last 200 million years.

The atmosphere is a mixture of gases that surround the Earth. The atmosphere gets thinner as you go further from the Earth's surface. It is difficult to say exactly where the atmosphere stops. The proportion of these gases in the atmosphere has remained much the same for the last 200 million years.

Air is the mixture of gases in the lower part of the atmosphere. About 99% of the air is made up of nitrogen and oxygen. About four-fifths of the air is nitrogen and one-fifth is oxygen.

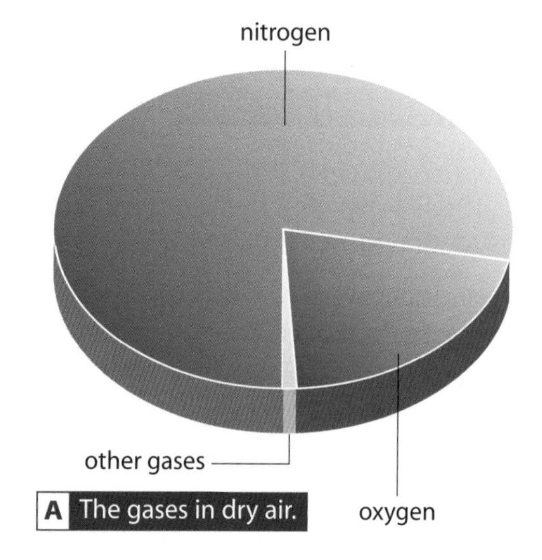

A The gases in dry air.

1 a What is the main gas in air?
b What fraction of air is this gas, roughly?

2 a What gas makes up most of the rest of air?
b What fraction of air is this gas, roughly?

3 How long has the atmosphere been like this?

There are small amounts of other gases which make up the other 1%. Some of these are **noble gases**, mainly argon. Noble gases are unreactive gases in Group 0 of the Periodic Table. A small amount of carbon dioxide is also found in air.

Water vapour is also present in the air. The amount of water vapour changes, and so it is not included when giving the composition of air.

There are also trace amounts of harmful gases in the air. These include sulfur dioxide, nitrogen oxide and carbon monoxide. These gases are formed in a number of ways. Some are made by natural processes such as volcanic activity, or from plants and animals decomposing. Some are made by human activities such as burning fossil fuels.

Gas	Formula	Percentage in dry air
Nitrogen	N_2	78
Oxygen	O_2	21
Argon	Ar	0.9
Carbon dioxide	CO_2	0.04
Other gases		traces

B

water (H_2O) traces

oxygen (O_2) 21%

carbon dioxide (CO_2) 0.04%

nitrogen (N_2) 78%

argon (Ar) 0.9%

C The gases in dry air.

4 a There are small amounts of noble gases in the air. What are noble gases?

b Name the main noble gas found in the air.

5 List two other naturally occurring gases found in the air.

6 Why is the proportion of gases in air given without any water?

7 Explain why we say that air is a mixture.

8 List the three main elements in the air.

9 Name a compound found in the air.

The gases in air can be separated. The main gas, nitrogen, is a very unreactive gas. It is used inside food packaging, such as crisp packets, rather than air, because the oxygen in it would allow respiring bacteria to make the food go off faster.

10 Nitrogen is used in food packaging. Explain why.

D Nitrogen is the gas in crisp packets.

11 Draw a bar chart to show the main gases in air.

The noble gases

By the end of this topic you should be able to:

- recall that the noble gases are the elements in Group 0 of the Periodic Table
- recall that the noble gases are found in air
- recall that the noble gases are all very unreactive
- describe some uses of the noble gases.

The noble gases are the elements in Group 0 of the Periodic Table. They are helium (He), neon (Ne), argon (Ar), krypton (Kr), xenon (Xe) and radon (Rn).

Group	1	2											3	4	5	6	7	0 or 8
				H														He
	Li	Be											B	C	N	O	F	Ne
	Na	Mg											Al	Si	P	S	Cl	Ar
	K	Ca	Sc	Ti	V	Cr	Mn	Fe	Co	Ni	Cu	Zn	Ga	Ge	As	Se	Br	Kr
	Rb	Sr	Y	Zr	Nb	Mo	Tc	Ru	Rh	Pd	Ag	Cd	In	Sn	Sb	Te	I	Xe
	Cs	Ba	La	Hf	Ta	W	Re	Os	Ir	Pt	Au	Hg	Tl	Pb	Bi	Po	At	Rn

A Group 0 in the Periodic Table.

Elements in the same group of the Periodic Table are similar. All the elements in Group 0:

- are gases at room temperature
- have very low melting and boiling points
- are non-metals
- are very unreactive (most of them do not react with anything)
- have particles that are just individual atoms.

Small amounts of the noble gases are found in air. The main noble gas in the air is argon (0.9%) with much smaller amounts of the other noble gases. The noble gases can be separated from the other gases in air by fractional distillation.

1 What are the noble gases?

2 Make a list of the noble gases.

3 Which is the main noble gas in the air?

4 The noble gases are elements. What are elements?

5 Why do the noble gases have similar properties?

6 Comment on the melting and boiling points of the noble gases.

Although they are very unreactive, noble gases do have a number of uses. They have some of these uses because they are very unreactive.

Use	Noble gas used	Reasons for use
Filament light bulbs	argon	Argon is the gas inside filament light bulbs. It is used because it does not react with anything. The temperature of the tungsten filament is over 2000 °C and at this temperature oxygen and nitrogen would react with the filament.
Electric discharge tubes	helium neon argon krypton xenon	Each gas glows a different colour if a large voltage is applied across it in a glass tube.
Airships and balloons	helium	Helium is used in airships and balloons because it is less dense than air. This means that the airship or balloon floats. Hydrogen was used in airships in the past but it is flammable. In 1937, the Hindenburg airship, the largest aircraft ever built, caught fire as it approached its landing at the end of its journey from Germany to the USA. Thirty-six people were killed. Helium is not flammable.

B

7 Explain why argon is used in filament light bulbs.

8 Explain why noble gases are used in electric discharge tubes.

9 Explain why helium is used in airships and balloons.

10 Draw a concept map to summarise the important facts about the noble gases.

P D P P

The early atmosphere

By the end of this topic you should be able to:

- describe which gases made up the early atmosphere
- explain where the gases in the early atmosphere came from
- explain how the atmosphere has slowly changed to form the one we have today.

The Earth is thought to be 4.6 billion years old. It is believed that the atmosphere that we have today is very different to the one the Earth used to have. However, many of the current theories are far from certain.

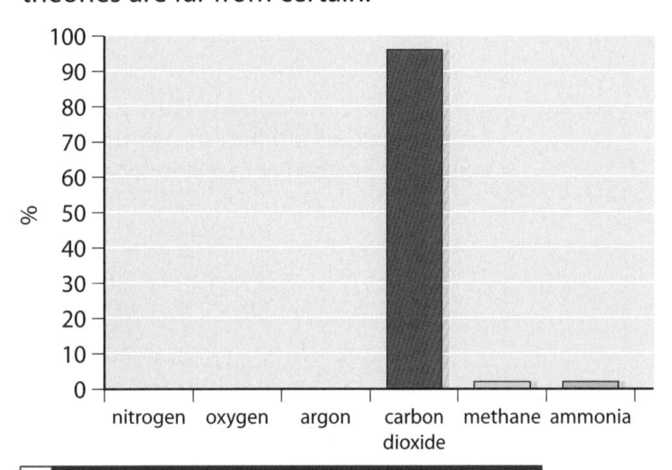

A The atmosphere when the Earth was young.

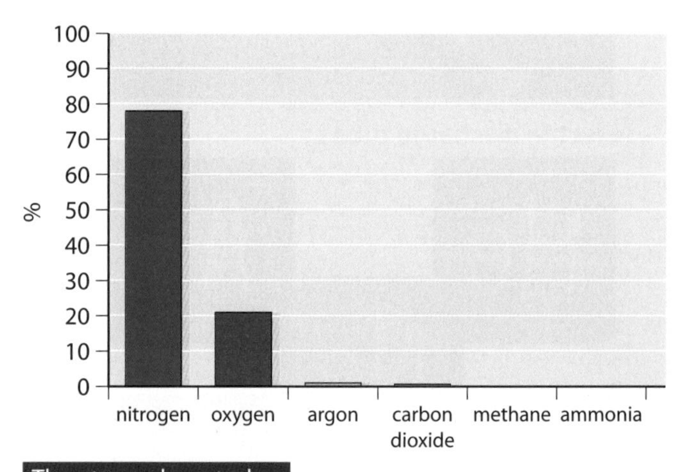

The atmosphere today.

It is believed that the gases which formed the early atmosphere came from inside the Earth. During the first billion years, the Earth was much hotter and there was a lot of volcanic activity. Volcanoes today release small amounts of gases which include carbon dioxide, water vapour, ammonia and methane. Scientists therefore believe that these gases would have been in the early atmosphere.

The Earth lies between Mars and Venus. Mars and Venus have very similar atmospheres containing mainly carbon dioxide (over 95%). It is believed that the Earth's early atmosphere was mainly carbon dioxide, with small amounts of water vapour, methane and ammonia. Scientists believe that the evolution of life on Earth caused our atmosphere to change. There is no life on Mars or Venus so their atmospheres have remained the same.

Oxygen is not released by volcanoes. This and other evidence leads most scientists to believe that there was little or no oxygen in the early atmosphere.

As the hot, young Earth cooled down, the water vapour in the air is thought to have condensed. This would have led to the formation of oceans.

The volcanoes gave out...

...lots of carbon dioxide.

...lots of steam.

The atmosphere also contained small amounts of methane and ammonia.

The steam condensed to make liquid water. This water made the oceans and seas.

B The gases in the early atmosphere came from volcanoes.

Photosynthesising organisms, including plants, evolved. They use up carbon dioxide and make oxygen. The oxygen we have in the atmosphere today is produced by plants. Over time, the amount of carbon dioxide decreased and the amount of oxygen increased. Most of the carbon from the carbon dioxide is now in fossil fuels or sedimentary rocks (mainly limestone).

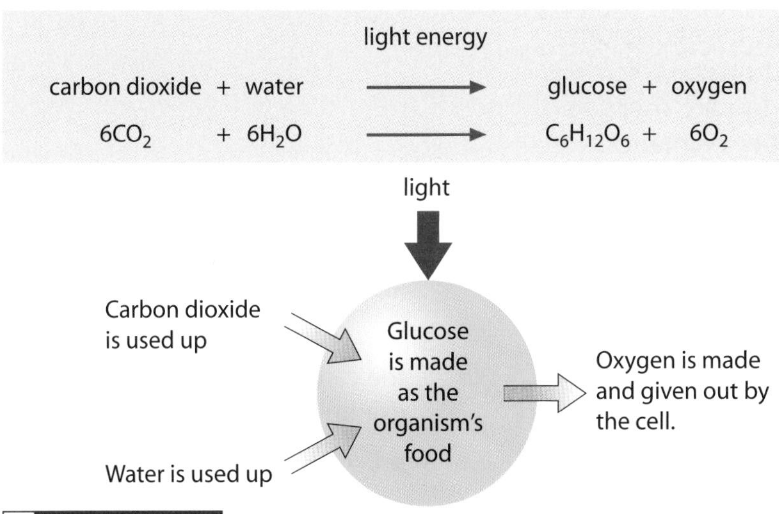

C Photosynthesis.

The ammonia in the early atmosphere would have reacted with the oxygen, producing some of the nitrogen in the atmosphere. However, most of the nitrogen was formed by bacteria from nitrogen-containing compounds in the soil. The methane also reacted with oxygen.

The proportions of the gases in the atmosphere are thought to have remained constant for the last 200 million years.

1 Where did the gases that formed the early atmosphere come from?

2 What gases are thought to have made up the early atmosphere?

3 How did the oceans form?

4 Why do scientists believe there was little or no oxygen in the early atmosphere?

5 What happened on Earth that changed the atmosphere to what it is today?

6 Describe how the oxygen in the atmosphere formed.

7 What process removed most of the carbon dioxide from the air?

8 What happened to the ammonia and the methane?

9 What else produced the nitrogen in the air?

10 Draw a flow chart to show the changes that have taken place in the atmosphere.

Carbon dioxide in the atmosphere

By the end of this topic you should be able to:

- recall that most of the carbon from the carbon dioxide in the early atmosphere is locked up in sedimentary rocks and fossil fuels
- recall that the amount of carbon dioxide in the atmosphere has remained constant for the last 200 million years
- explain the effects of human activities on the atmosphere.

When the Earth was young, the main gas in the atmosphere is thought to have been carbon dioxide, but there is very little now. Much of it was removed by plants during photosynthesis. The carbon from the carbon dioxide became part of the plant structures. When the plants died and decayed, they formed fossil fuels such as coal.

As the Earth cooled, water vapour in the air condensed to form the oceans. About half of the carbon dioxide from the atmosphere is thought to have dissolved in the oceans. Some of this was used by sea plants and algae during photosynthesis. Some of these creatures also died and decayed, locking carbon away as crude oil and natural gas.

Some carbon dioxide reacted with substances in sea water. This formed some insoluble compounds, such as calcium carbonate, that sank to the bottom as sediment. It also formed soluble compounds, such as calcium hydrogencarbonate, which was used by sea creatures to make calcium carbonate shells. As these creatures died, their shells fell into the sediment. Over time, these sediments formed sedimentary rocks such as limestone, locking away the carbon in rocks.

B Much of the carbon is locked up in sedimentary rocks like limestone.

A Much of the Earth's carbon is locked up in fossil fuels.

1 What was the main gas in the air when the Earth was young?

2 How much carbon dioxide is in the air now?

3 Where is most of the carbon from the carbon dioxide now? Give two examples.

4 a Explain how plants remove carbon dioxide from the air.
b Explain how this carbon ended up in fossil fuels.

5 a Explain how the oceans removed carbon dioxide from the air.
b Explain how some of this carbon ended up in
(i) fossil fuels
(ii) sedimentary rocks.

Processes that absorb and release carbon dioxide from the air have been in balance for the last 200 million years or so, therefore the amount of carbon dioxide has remained at around 0.03%.

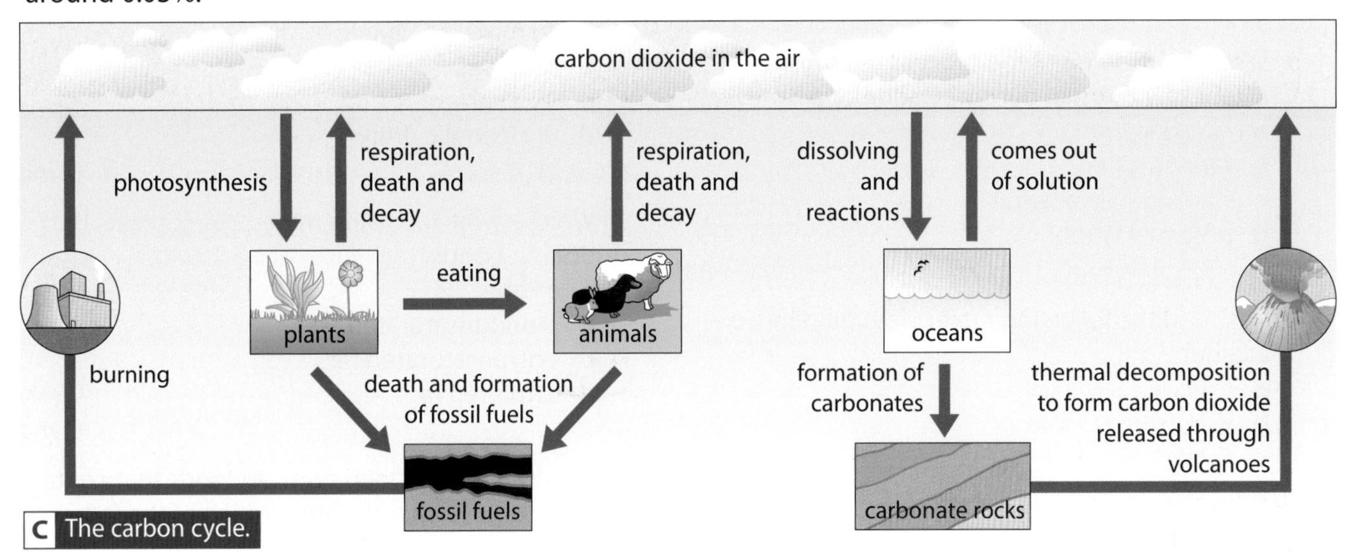

C The carbon cycle.

Increase in carbon dioxide

In recent years carbon dioxide levels have started to rise. They have risen to nearer 0.04%.

It is thought that some of this rise is due to large-scale burning of fossil fuels. When fossil fuels are burned they produce carbon dioxide.

6 Give four ways in which carbon dioxide is released into the atmosphere.

7 Why has the amount of carbon dioxide in the atmosphere remained constant for the last 200 million years?

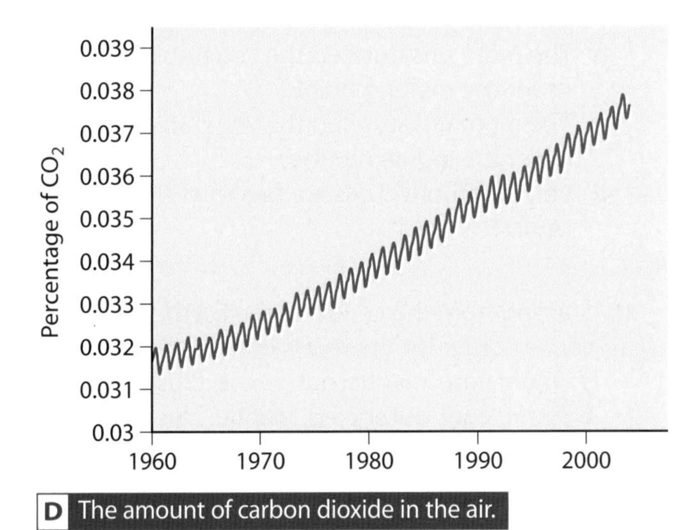

D The amount of carbon dioxide in the air.

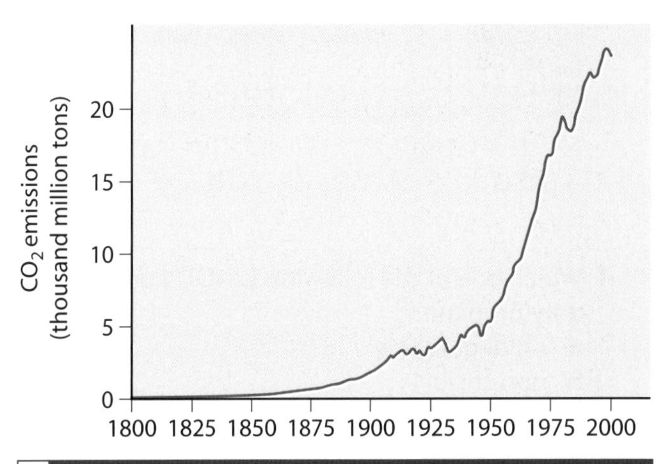

E The amount of carbon dioxide released into the atmosphere from burning fossil fuels.

Large-scale deforestation is also thought to be contributing to the increase in carbon dioxide.

8 Why is the amount of carbon dioxide in the air increasing? Give two reasons.

9 Why is this thought to be a problem?

10 Draw a table with two columns, one for processes that remove carbon dioxide from the air and one for processes that release it into the air. Complete the table to summarise this topic.

Assessment exercises

Part A

1 What is the name of the process that produces alkenes by thermal decomposition of alkanes?

a cracking
b fermentation
c fractional distillation
d polymerisation

(1 mark)

2 Which of the following molecules could be an alkene?

a C_3H_8
b C_2H_6
c C_6H_{12}
d C_4H_{10}

(1 mark)

3 Ethene (C_2H_4) is used to make poly(ethene). What is the correct structure of poly(ethene)?

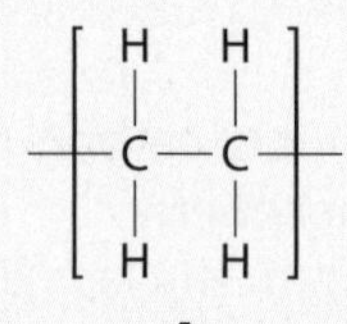
A

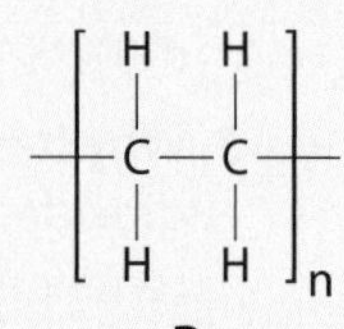
B

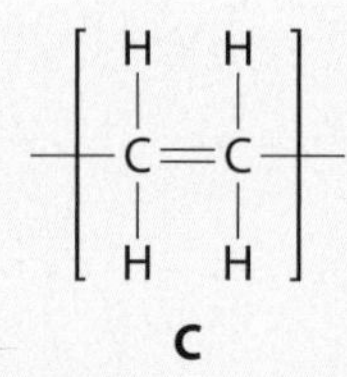
C

D

(1 mark)

4 Which one of the following is NOT a property of poly(propene)?

a biodegradable
b non-toxic
c strong
d unreactive

(1 mark)

5 Ethanol can be made from ethene by reaction with steam at high temperature. The reaction will only take place in the presence of:

a an alkane.
b a catalyst.
c an enzyme.
d a hydrocarbon.

(1 mark)

6 Which of the following is NOT a use for ethanol?

a To make polymers.
b As a fuel.
c As a solvent.
d In alcoholic drinks.

(1 mark)

7 Which of the following molecules contains NO C=C double bonds?

a alkene
b monounsaturated fat
c polyunsaturated fat
d saturated fat

(1 mark)

8 Emulsifiers are added to many foods that contain mixtures of oils and water. Which of the following is NOT a reason for adding an emulsifier?

a To allow the oil and water to mix.
b To improve the texture of the food.
c To saturate the oils.
d To thicken the mixture.

(1 mark)

9 Which of the following is NOT true about vegetable oils?

a Vegetable oils can be extracted from many nuts, fruits and seeds.
b The more unsaturated the vegetable oil, the lower the melting point.
c The more unsaturated the vegetable oil, the lower the iodine number.
d Polyunsaturated oils are healthier than saturated ones

(1 mark)

10 Starting from the centre of the Earth, what is the correct order for the layers of the Earth?

a inner core, mantle, outer core, crust
b inner core, outer core, mantle, crust
c outer core, inner core, mantle, crust
d outer core, mantle, crust, inner core,

(1 mark)

11 How did scientists believe that mountains were formed before Alfred Wegener first suggested the idea of continental shift?

a By the shrinking of the Earth's crust.
b Less dense rocks rose up above more dense rocks.
c Melting of rocks in the crust.
d The Earth's mantle expanded.

(1 mark)

12 Which of the following shows the gases in the present atmosphere in order of abundance – starting with the most abundant gas?

a oxygen, carbon dioxide, nitrogen, argon
b oxygen, nitrogen, methane, carbon dioxide
c nitrogen, oxygen, argon, carbon dioxide
d nitrogen, oxygen, carbon dioxide, argon

(1 mark)

13 Argon is in Group 0 of the Periodic Table. Which of the following is NOT true about argon.

a It is unreactive.
b It is used in balloons because it is less dense than air.
c It is used in electric discharge tubes.
d It is used in filament light bulbs.

(1 mark)

14 In which of the following processes is carbon dioxide NOT removed from the atmosphere?

a Gases dissolving in the oceans.
b Gases reacting with substances in the oceans.
c Respiration in plants.
d Photosynthesis in plants.

(1 mark)

15 Which of the following is NOT a sign that a volcano is likely to erupt?

a Gases being released from the volcano.
b Ground movement.
c An increase in sulphur dioxide emissions.
d An increase in temperature.

(1 mark)

Total (Part A) 15 marks

Part B

1 The polymer poly(propene) is made from propene. Propene is made by thermal decomposition of naphtha which is a fraction of crude oil. The structure of propene is shown.

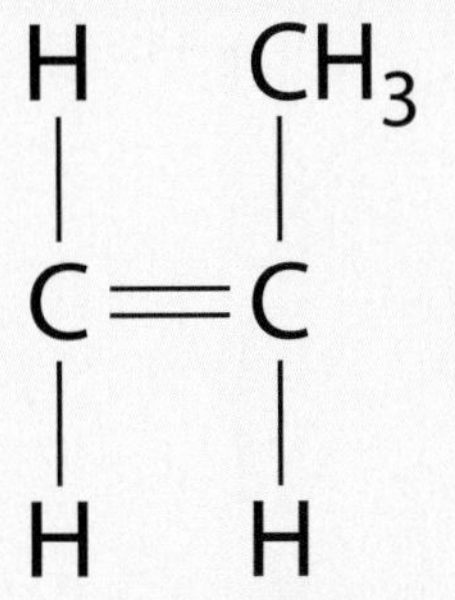

a What is the name of the thermal decomposition reaction in which propene is formed? *(1 mark)*
b Give the formula of propene. *(1 mark)*
c Write an equation for the formation of poly(propene) from propene. *(2 marks)*
d What is a polymer? *(2 marks)*
e Poly(propene) is non-biodegradable. What does this mean? *(2 marks)*
f Waste poly(propene) is often buried in landfill. Give two other ways in which it is disposed of. *(2 marks)*

2 Rapeseed oil is the commonest vegetable oil in the UK. It contains some unsaturated oils.

a Describe a test to show that rapeseed oil contains unsaturated oils. *(2 marks)*
b What must be added to rapeseed oil for it to mix with water? *(1 mark)*
c Rapeseed oil needs to be hardened to make margarine.
  i What is the rapeseed oil reacted with? *(1 mark)*
  ii Under what conditions is this reaction carried out? *(2 marks)*

3 In 1915, Alfred Wegener first published his ideas about continental drift. At first they were rejected, but in the 1960's they became the basis for the modern theory of plate tectonics.

a Give two reasons why his ideas were rejected by most scientists at the time. *(2 marks)*
b Give two pieces of evidence that support Wegener's ideas. *(2 marks)*

4 These pie charts show the composition of the atmosphere at different times in the history of the Earth.

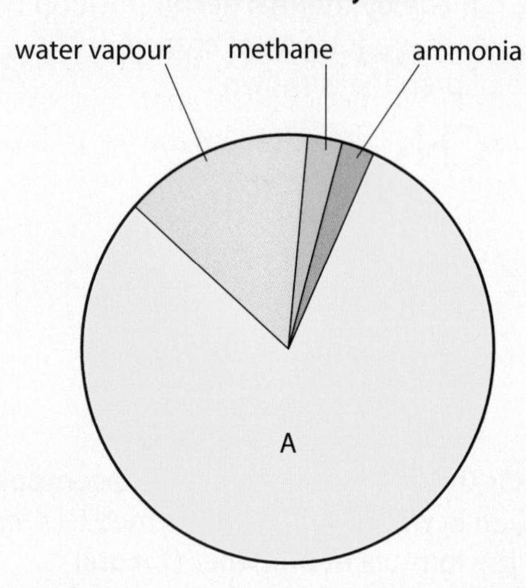

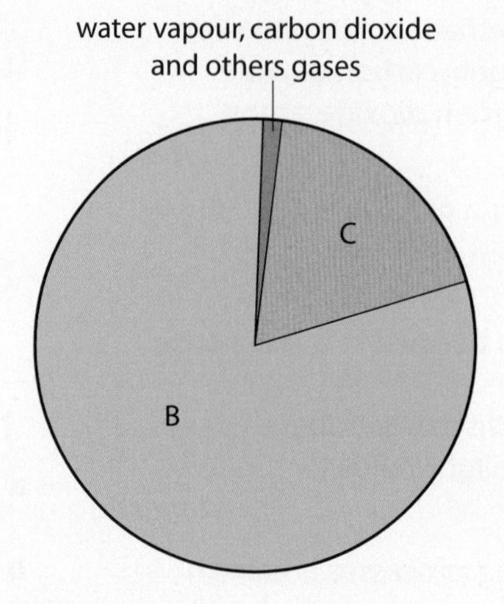

a What are the gases labelled **A**, **B** and **C**? *(3 marks)*

b Where did the gases in the early atmosphere come from? *(2 marks)*

c **i** Give two ways in which carbon dioxide is removed from the atmosphere. *(2 marks)*

ii Give two ways in which carbon dioxide is returned to the atmosphere. *(2 marks)*

iii Why are carbon dioxide levels increasing? *(1 mark)*

Total (Part B) 30 marks

Investigative Skills Assessment

Vegetable oils are unsaturated. This means that they contain C=C double bonds. The more C=C double bonds they contain, the more healthy (or less unhealthy) the oil. Hasina carried out an investigation to compare several different vegetable oils and place them in order of unsaturation.

Hasina added bromine water to samples of four different oils. Bromine water reacts with unsaturated molecules, losing its yellow-orange colour and becoming colourless. The more bromine water that reacts with a vegetable oil, the more C=C double bonds and so the more unsaturated the oil.

The results of the investigation are shown in table A, below.

Oil	**Volume of bromine water reacting with oil sample (cm^3)**			
	Experiments			**Average**
	1	**2**	**3**	
Sunflower oil	44	51	46	47
Olive oil	29	25	30	28
Rapeseed oil	41	33	40	38
Corn oil	57	42	45	48

A Investigation results.

1 Copy and complete this statement about the investigation. Hasina was trying to find out… (*1 mark*)

2 a What was the independent variable? (*1 mark*)
  b Is this variable best described as a categoric, ordered, continuous or discrete variable? (*1 mark*)

3 a What was the dependent variable? (*1 mark*)
  b Is this variable best described as a categoric, ordered, continuous or discrete variable? (*1 mark*)

4 State two variables that were controlled to make this a fair test. (*2 marks*)

5 a Name a different piece of apparatus that could have been used to measure out the bromine water in each experiment. (*1 mark*)
  b Suggest why the burette was the best piece of apparatus to use. (*1 mark*)
  c Suggest why Hasina added bromine water 2 cm^3 at a time rather in bigger amounts such as 10 cm^3. (*1 mark*)

6 Why do you think there was some variation in the results for each oil? (*1 mark*)

7 ✎ Describe how you would carry out the experiment yourself. Explain the reasons for your suggestions. (*4 marks*)

8 Are there any results in the table that do not seem to fit in with the others (anomalous results)? If so, which results are they? (*1 mark*)

9 Hasina's teacher carried out a trial experiment to decide how much oil should be used. Suggest how this was done. (*1 mark*)

(Total = 17 marks)

Glossary

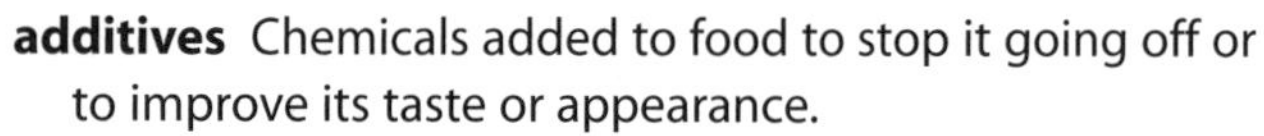

D

additives Chemicals added to food to stop it going off or to improve its taste or appearance.
air The mixture of gases in the lower part of the atmosphere.
alkane Hydrocarbon with the general formula C_nH_{2n+2}.
alkene Unsaturated hydrocarbon containing one C=C double bond.
biodegradable Can be broken down by microbes.
biodiesel Diesel made from vegetable oils.
carbon neutral A fuel that releases the same amount of carbon dioxide when it burns as the crops from which it is made took in as they grew.
catalyst Chemical that speeds up a reaction but does not get used up.
chromatography Separation of substances by their solubility in a solvent which is moving through a solid (e.g. paper).
continental drift The movement of continents round the world.
convection current A flow of liquid or gas caused by part of it being hotter or colder than the rest.
cracking Breakdown of long alkanes into shorter alkanes and alkenes.
crust The outer part of the Earth.
decompose Break down into simpler substances.
distillation A method used to separate the solvent from a solution.
earthquake When the ground shakes due to the sudden movement of tectonic plates.
emulsion Mixture of an oil and water in which tiny droplets of one substance are spread out through the other substance.
emulsifier Substance added to two liquids that do not mix to form an emulsion.
E-number Number given to food additives that have passed safety checks by the European Food Safety Authority.
enzyme Protein molecule that speeds up reactions in living cells.
fraction Mixture of chemicals with similar boiling points produced by fractional distillation.
fractional distillation Used to separate two or more liquids with different boiling points.
hydrocarbon Compound containing hydrogen and carbon only.
hydrogel Polymers that can absorb a lot of water.
hydrogenation Addition of hydrogen.
incinerator A furnace for burning waste under controlled conditions.
landfill The burying of large amounts of rubbish in the ground.
lithosphere The Earth's crust and the upper part of the mantle.
mantle The layer inside the Earth between the crust and the core.
monomer Short molecule used to make a polymer.
monounsaturated Molecule containing one C=C double bond.
noble gases Unreactive gases in Group 0 of the Periodic Table.
non-biodegradable Cannot be broken down by microbes.
non-renewable resource A resource that cannot be replaced once it has been used. Non-renewable resources will eventually run out.
poly(ethene) Polymer made from ethene.
polymer Long chain molecule made from joining many short molecules (monomers) together.
polymerisation Reaction in which many short molecules (monomers) are joined together to make a long molecule (polymer).
polyunsaturated Molecule containing more than one C=C double bond.
recycle Reuse.
renewable A resource that can be replaced once it has been used (e.g. plants).
saturated Molecule containing no C=C double bonds.
shape memory polymer Polymer that changes shape as the temperature changes, but can change back to its original shape again.
slime A thick, sticky, slippery substance.
smart materials Materials that have one or more properties that change due to changes in conditions such as temperature, light, pH, electric or magnetic fields.
solvent A chemical used to dissolve something.
tectonic plates A large section of the Earth's lithosphere.
thermal decomposition Breakdown of a substance into simpler substances using heat.
trans fats Unsaturated fats produced by the hydrogenation of vegetable oils.
unsaturated Molecule containing one or more C=C double bonds.
viscosity How easily a liquid flows.
volcano An opening in the Earth's crust where magma can escape from inside the Earth to the surface.

Energy and electricity

We need energy for everything we do. The source of most of the energy we use comes from burning fossil fuels. This energy has to be paid for, and burning fossil fuels adds carbon dioxide to the atmosphere. The carbon dioxide is contributing to environmental problems, including global warming. This means it is important to try to reduce the amount of energy we use.

A The Hockerton Housing Project (HHP) houses.

B

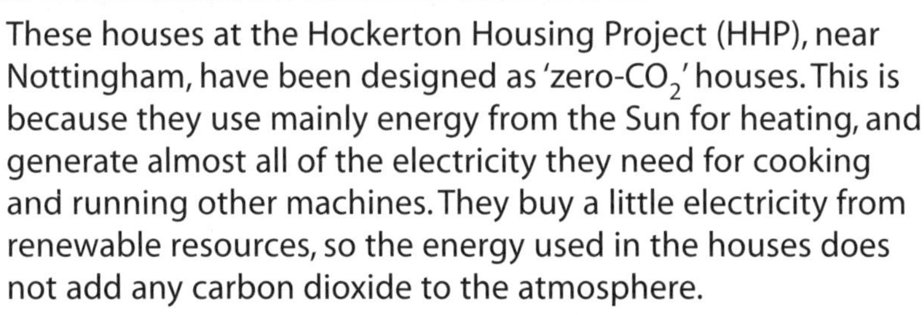

These houses at the Hockerton Housing Project (HHP), near Nottingham, have been designed as 'zero-CO_2' houses. This is because they use mainly energy from the Sun for heating, and generate almost all of the electricity they need for cooking and running other machines. They buy a little electricity from renewable resources, so the energy used in the houses does not add any carbon dioxide to the atmosphere.

By the end of this unit you should:

- be able to identify and explain the different ways in which heat is transferred
- be able to evaluate ways of reducing heat transfers
- know what is meant by the efficient use of energy
- be able to explain why electrical devices are so useful
- be able to describe how electricity is generated and supplied to our homes
- be able to discuss the advantages and disadvantages of different ways of generating electricity.

1 Write down all the different sources of energy you use in one day.

2 **a** Look at the photographs of the HHP houses. Make a list of the differences between these houses and rows of houses in your area.
b Which of these differences help to save energy in the HHP houses?
c Why is it important to build homes that use as little energy as possible for heating?
d What energy resources are used to generate electricity for your home?

3 Make a list of the things you need to find out to help you to give better answers to the questions above.

Conducting heat

By the end of this topic you should be able to:

- describe how heat energy is transferred through solid materials
- list materials that are thermal conductors and thermal insulators
- explain why different materials are used in a house.

If an object is warmer than its surroundings, heat will be transferred from the object to the surroundings until they are both at the same temperature.

A Hot and cold drinks.

1 Look at photograph A.
  a What will happen to the temperature of each drink?
  b Explain your answers.

One of the reasons the HHP houses do not need boilers and radiators for heating is that they are very well **insulated**. Materials used in the walls are **thermal insulators** that do not allow heat to flow through them easily. Insulation in the walls reduces the amount of heat energy that is transferred from the warm rooms to the outside.

Most materials made from **non-metals** are thermal insulators. Materials that contain a lot of trapped air are usually very good insulators.

2 How does insulation help to keep the HHP houses cool in the summer?

3 Look at the insulating materials in photograph B.
  a Which one do you think is the best thermal insulator?
  b Explain your answer.

loft insulation

polystyrene foam

bubble wrap

B Insulating materials.

Heat can travel in different ways. Heat travels through solid materials by **conduction**. The **particles** that make up solids are held closely together in fixed positions. The particles cannot move around, but they can vibrate. The higher the temperature, the more the particles vibrate.

Metals are good **thermal conductors**. When one end of a piece of metal is heated, the particles in the heated part start to vibrate more. These particles bump into nearby particles and make them vibrate more, and so on. The heat energy is conducted through the metal.

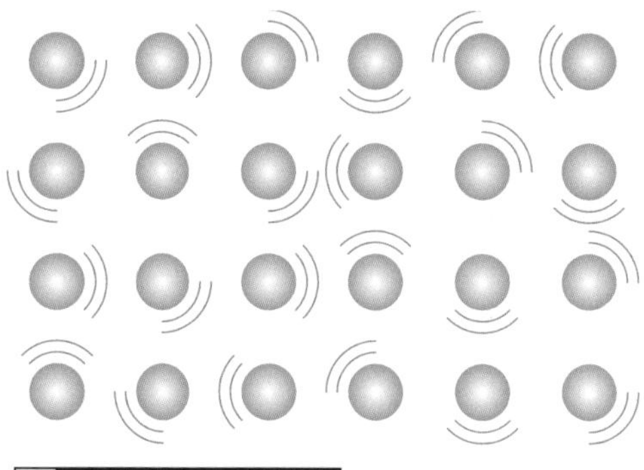

C Particles in a solid.

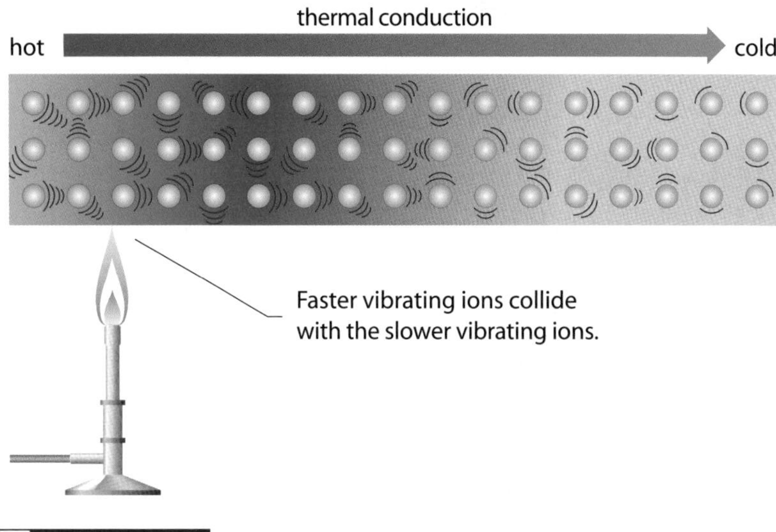

D Heating metal.

4 **a** How are particles arranged in a solid?
  b How are particles arranged in a gas?

5 Why do you think gases are not good thermal conductors?

6 Which properties of metals are the most important for the following uses:
  a radiator
  b saucepan
  c taps
  d wires and switches.

Choosing materials

Houses and the things in them are made from many different materials. The materials are chosen because of their properties. Different **properties** are important for different jobs.

Material	Properties
Metal	thermal conductor, electrical conductor, hard, strong, high melting point
Plastic	thermal insulator, electrical insulator, easy to mould into shape, low melting point, cheap, does not break easily
Glass	thermal insulator, electrical insulator, transparent, breaks easily

E Properties of different materials.

7 Choose a material for each of the following uses. Explain your choices.
  a the outside of a plug
  b a window
  c a saucepan handle
  d a bowl for mixing food.

8 The conservatory on the HHP houses has double-glazed windows. These have two sheets of glass with an air gap between them.
  a Why do you think double glazing provides better insulation than single glazing?
  b Explain why new houses are fitted with double-glazing.

9 Explain how heat energy is transferred through a saucepan to the food inside it.

Keeping warm

By the end of this topic you should be able to:

- describe how the rate of heat transfer depends on temperature difference
- describe different types of variables.

Choosing the right insulating materials is important in building houses that do not need much energy for heating. It is also important to use the correct amount of material.

A The insulation used in the HHP houses is much thicker than the insulation used in standard housing.

Less heat is lost if thicker insulation is used, but thicker insulation costs more and takes up more space. An architect needs to work out the best thickness for a particular house.

1 Why do the HHP houses use less energy for heating than standard houses?

2 What problems might there be if a house was built with:
a not enough insulation
b too much insulation?

Testing insulation

When you investigate a question scientifically you measure variables. If you are testing insulation, you might do several tests with different thicknesses. The thickness of the insulation is the independent variable (the variable you change). You could measure the temperature of something wrapped in the insulation. The temperature is the dependent variable. There are other variables, such as the type of insulation, which you have to keep the same for each test to make sure your test is fair. These are control variables.

B Testing insulation.

Variables can be of different types. For example, if you count the number of layers of the insulation you are testing, your independent variable will also be a discrete variable.

Variable	Description	Example
Categoric	labels	type of insulation
Ordered	descriptions you can put into order	thin, medium or thick insulation
Discrete	whole numbers	1, 2 or 3 layers of insulation
Continuous	precise numbers (including decimals)	measurements of the thickness of the insulation

C Different types of variable.

3 Look at photograph B.
  a What type of independent variable is being tested? Explain your answer.
  b How could the independent variable be made into a continuous variable?

Changing the design

You have to supply more energy to keep a house warm on a cold winter day than on a mild spring day. This is because a house loses heat faster when there is a bigger temperature difference between the house and its surroundings.

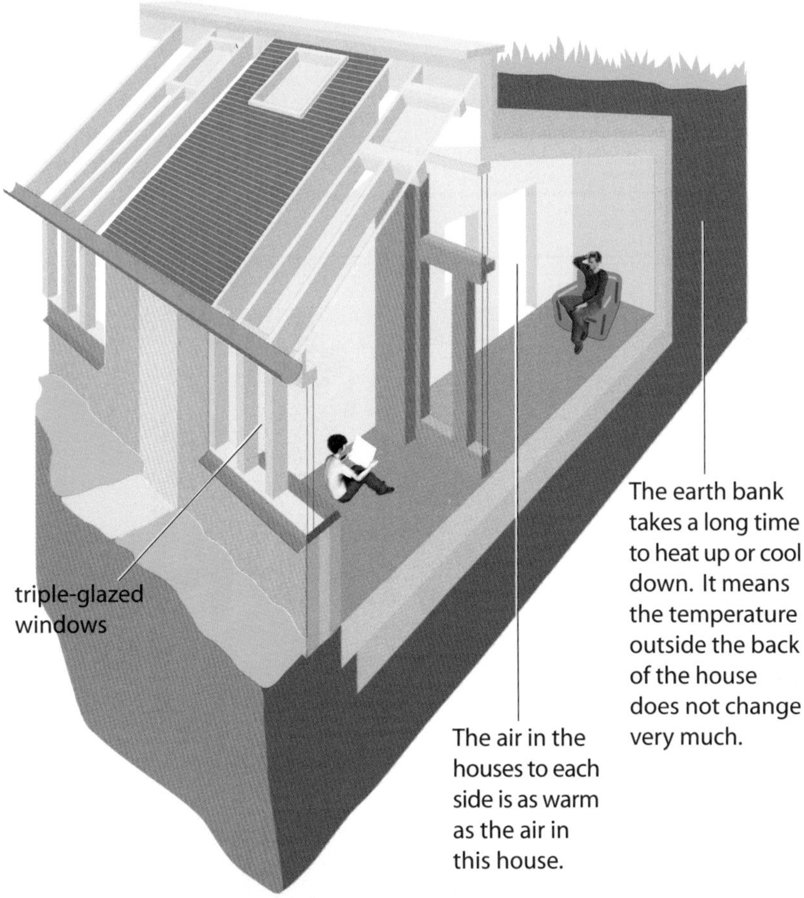

D The design of the HHP houses reduces the rate that heat is transferred out of the building in winter.

4 Look at diagram D. How does the earth bank help to keep the houses:
  a warm in winter
  b cool in summer?

5 Why is it cheaper to heat a terraced house than a detached house?

6 Why do you think triple-glazed windows are better insulators than double-glazed windows?

7 You are going to ask an architect to design a house for you.
  a Write a list of features you want her to include in the design to make your heating bills as low as possible.
  b Explain why each feature is needed to keep the costs down.

Convection

By the end of this topic you should be able to:

- explain how heat is transferred by convection
- describe some uses of convection
- explain why the shape of something can affect how quickly it transfers heat.

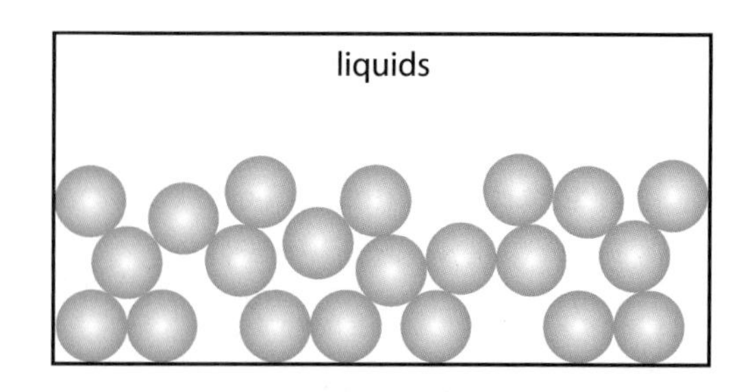

In liquids the particles are close together but they can move past each other.

Fluids (liquids and gases) are poor conductors of heat, which is why insulating materials such as polystyrene foam contain pockets of trapped air. However, air and other fluids can transfer heat energy if they are free to move. This is called **convection**.

1 What is convection?

2 Why doesn't convection happen in solids?

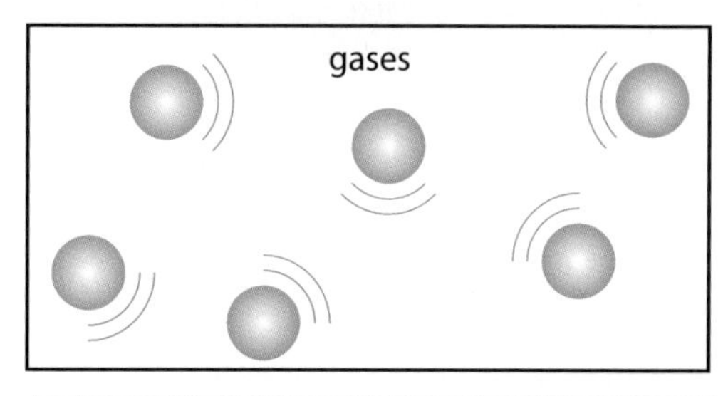

A In gases the particles are far apart and can move around.

When a fluid is heated the particles move around faster and take up more space. This makes the fluid less **dense**, so it floats up through the colder fluid around it. This sets up a flow called a **convection current**. Diagram B explains how a heater on the side of a fish tank can heat all of the water in the tank.

B Convection in a fluid.

3 Why do fluids get less dense when they are heated?

4 Why is air only a good insulator if it cannot move?

Convection currents can form around any object that is warmer or cooler than its surroundings. For example, heat energy from a mug of hot tea will be transferred to the air around it, and this air will warm up and rise. Cooler air flows in to take its place, and this makes the tea cool faster. The tea would cool even faster if it was poured into a bowl, because there would be a bigger surface area where energy could be transferred from the tea to the air.

Heating and cooling in homes often depends on convection currents.

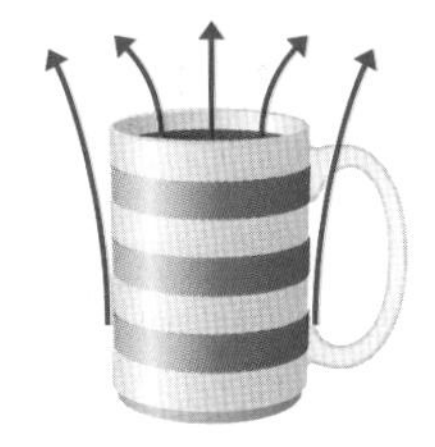

Heat is transferred to the air by the hot mug and the surface of the tea.

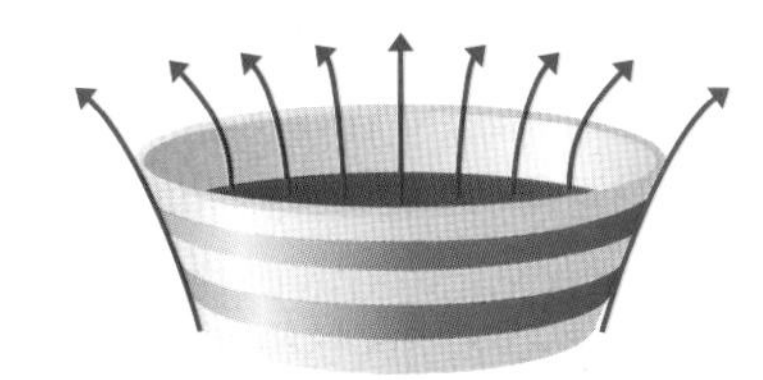

The tea and the bowl together have a bigger surface area than the mug, so heat can be transferred to the air faster.

C The tea in the bowl will cool faster than the tea in the mug.

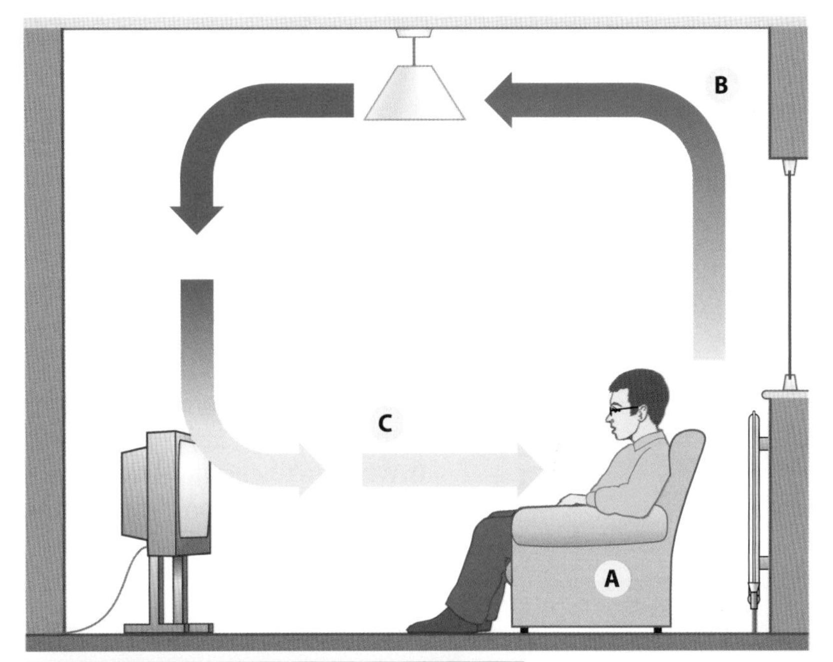

D A radiator can heat all of the air in a room.

E Convection currents help to keep the HHP houses cool in summer.

5 If the heating is turned off on a day in winter, why is the coldest place in a room likely to be under a window?

6 How do convection currents help to keep the house in photograph E cool?

7 **a** Why is the cooling element in a freezer at the top?
b Explain why the coldest part of a fridge is usually at the bottom.

8 Why are smoke detectors fitted to ceilings rather than walls?

9 Look at diagram D. Explain how the radiator can heat the whole room. Refer to what is happening at points A, B and C in your answer.

A V P D P P

Radiation

By the end of this topic you should be able to:

- recall that infra red radiation is the transfer of energy by electromagnetic waves
- explain that all bodies emit and absorb infra red radiation, and that hotter bodies radiate more energy
- describe which surfaces are best and worst at absorbing and emitting radiation.

Heat energy can travel through solids by conduction and through fluids by convection. Both of these need particles to pass on the energy. There is no matter in space between the Sun and the Earth, so heat energy from the Sun must reach us by some other means.

Heat energy can travel through space and through transparent materials as **infra red** or **thermal radiation**. These are a form of **electromagnetic wave**, similar to light waves.

1 a How does heat energy travel from the Sun to the Earth?
b Why can't heat travel from the Sun by convection or conduction?

A These **solar panels** absorb heat energy from the Sun to heat water.

Everything **emits** (gives out) and **absorbs** (takes in) infra red radiation. If two objects are the same size and shape, the hotter one will emit more energy than the cooler one.

2 Look at photograph B. What is the photograph showing?

3 The different colours in photograph B represent different amounts of infra red radiation being emitted.
a Which colour represents the hottest areas?
b Which colour represents the coolest areas?
c Explain how you worked out your answers to parts **a** and **b**.

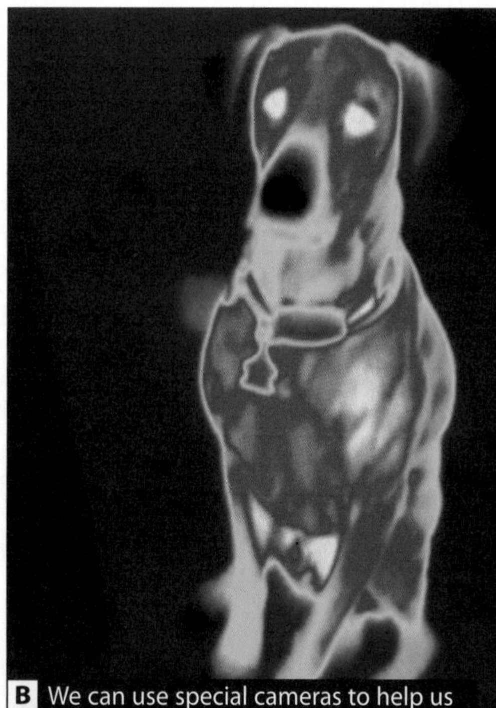

B We can use special cameras to help us to see the amount of infra red radiation emitted by different objects.

Infra red radiation can also be reflected. Light coloured, shiny surfaces are good at **reflecting** radiation. Dark, **matt** surfaces are good at absorbing radiation. Surfaces which are good at absorbing radiation are also good at emitting it.

4 Copy and complete table D using the words 'black' and 'white'.

	Best	**Worst**
Emitting infra red radiation		
Reflecting infra red radiation		
Absorbing infra red radiation		

D

Black surfaces are good at absorbing radiation.

White and silver surfaces are good at reflecting radiation.

C Absorption and reflection of infra red radiation.

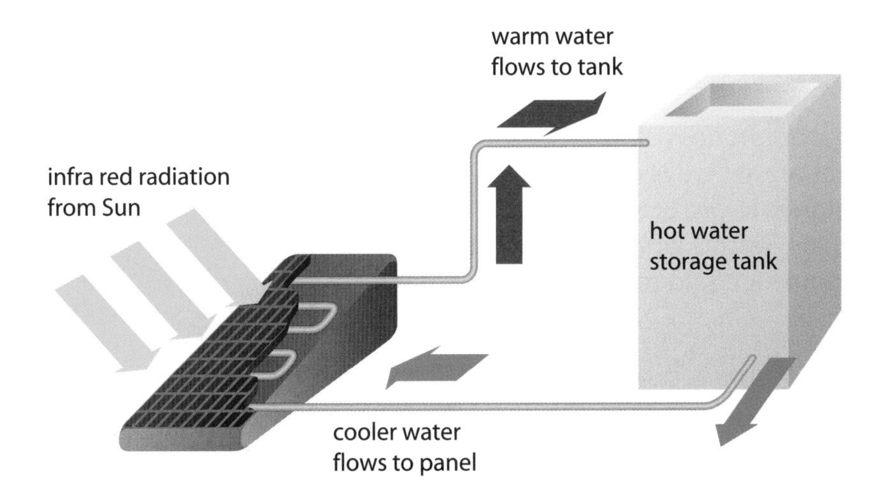

E Heating water with infra red radiation.

5 You can buy insulated mugs to keep hot drinks hot.
- **a** What is the best colour for an insulated mug?
- **b** Explain your answer.

6 Look at diagram E.
- **a** Why are the solar panels black?
- **b** Why doesn't the system need a pump to move the water round?
- **c** Which direction should the panels face? Explain your answer.

F This is a design for a house that will stay cool in the summer.

7 Look at the house in diagram F. Explain how each of the labelled features helps the house to stay cool.

8
- **a** Describe the differences between convection, conduction and radiation.
- **b** How can you change the amount of energy emitted or absorbed by an object?

Accurate measurements

HSW *By the end of this topic you should be able to:*

- explain why standard testing procedures are used
- recall some things to consider when choosing a measuring instrument
- describe ways to make sure that results are reliable.

The Government sets minimum standards of insulation for all new homes built in the UK. Architects need to calculate the amount of heat energy that will be lost through different parts of a building to make sure their design meets these standards. This means they need to know how much heat energy is transferred through the different types of material they could use.

Different building materials are tested in laboratories using standard procedures. Standard procedures are set ways of testing things, and allow different laboratories to get the same results if they are testing the same materials. Results which are the same when measured by different people are said to be reproducible.

1 a Why do building materials need to be tested?
b Why do testing laboratories follow standard procedures?

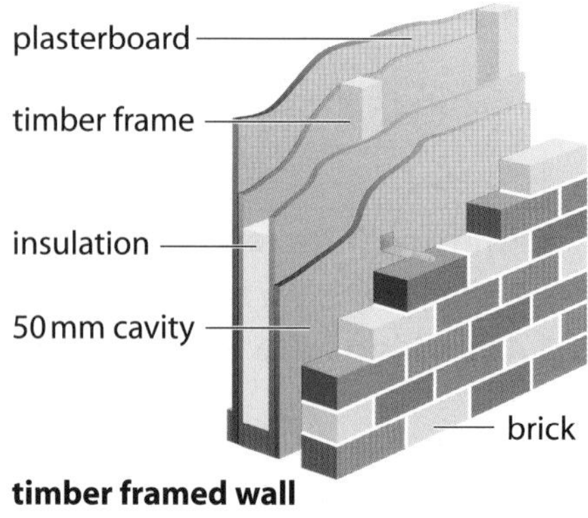

timber framed wall

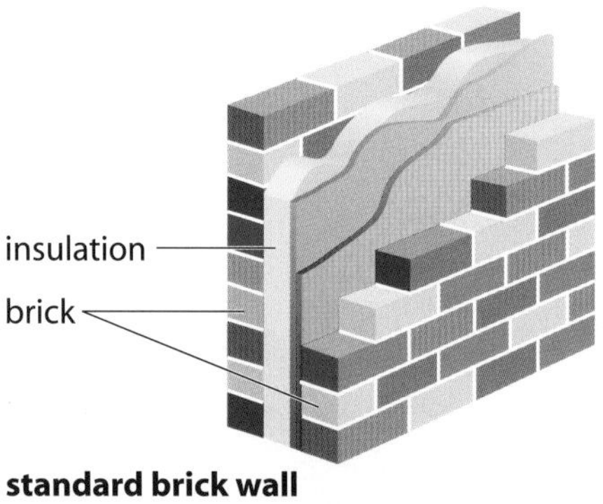

standard brick wall

A Which wall will provide the best insulation?

Testing laboratories must choose their instruments carefully to make sure their results are correct. For instance the instruments must be:

- accurate (they give readings close to the real value)
- sensitive enough (they can detect small enough changes)
- precise and reliable (they always give the same reading when measuring the same thing).

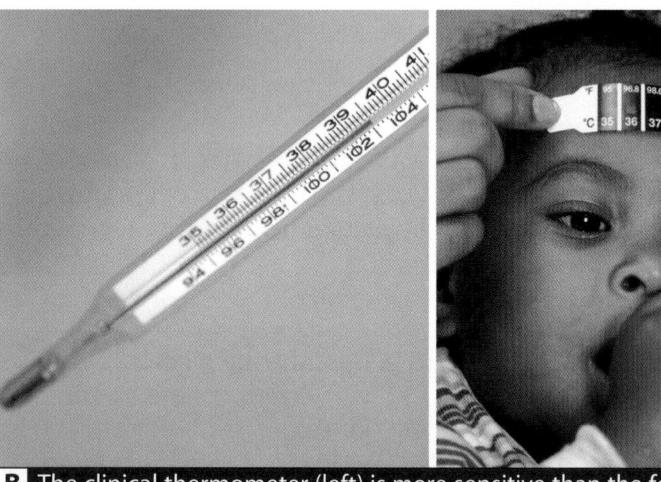

B The clinical thermometer (left) is more sensitive than the forehead thermometer (right), but only if it is used properly. It must be used in a standard way to give reliable results.

2 Which is more sensitive, a school electronic balance or bathroom scales? Explain your answer.

3 Why is it important that a clinical thermometer is:
a accurate
b sensitive
c reliable?

Even if you use a very good measuring instrument, there could still be errors in the results. You may have set up the instrument incorrectly, you may have made a mistake in reading it, or conditions in the laboratory may have changed.

To get the most accurate results, measurements are usually repeated several times. If there are any anomalous measurements (measurements that do not fit the pattern) you should try to find out why they are different. Any poor measurements should be ignored.

Once you have checked all the measurements, you can work out a mean and a range (the difference between the maximum and minimum values). The range will give you an idea of how precise your results are – the smaller the range the more precise the results.

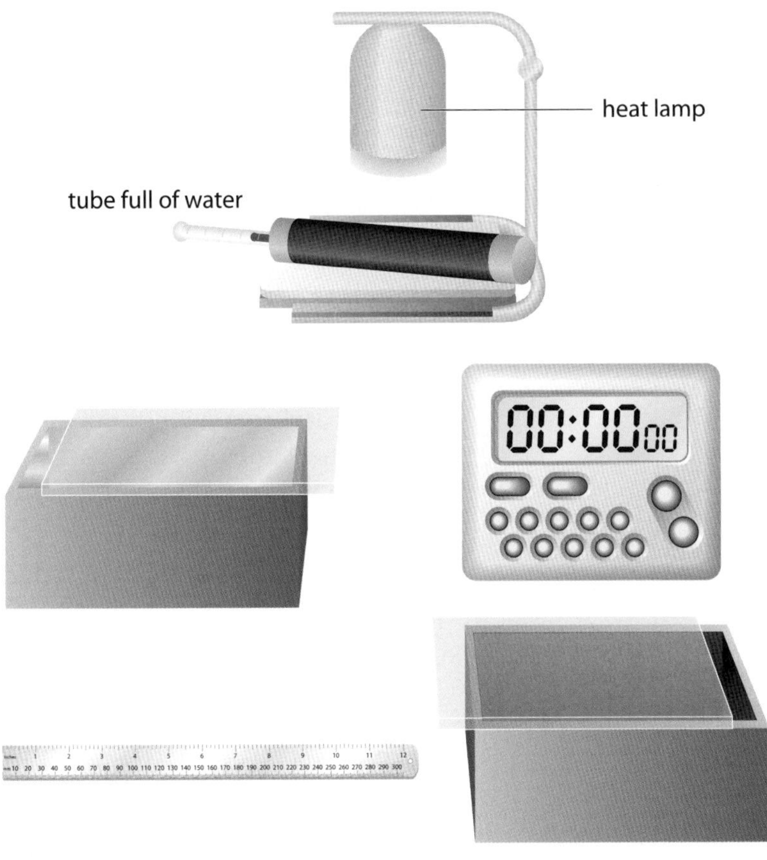

C Investigating designs for a solar panel.

Design	Temperature of water after 10 minutes (°C)			
	1st test	2nd test	3rd test	4th test
Tube only	25	27	33	24
Tube and silver box	43	34	44	41
Tube and black box	35	33	36	33

D Investigation results.

4 Look at diagram C. Sam is testing different designs for a solar panel to heat water.
  - **a** Suggest why Sam is going to test the different designs in the laboratory instead of outside in the sunshine.
  - **b** Which variables must Sam keep the same to make his test fair?

5 Table D shows the results of Sam's investigation.
  - **a** Which results are anomalous?
  - **b** Suggest what could have caused each of these two results to be different to the rest.
  - **c** Ignoring the anomalous results, calculate a mean and range for each design.

6 Write a conclusion for Sam's experiment.

7 **a** Explain what you have to think about when choosing a measuring instrument.
  - **b** How could you make sure the results of an investigation are as reliable as possible?

P D P P

P1a.6

Reducing heat transfers

By the end of this topic you should be able to:

- evaluate ways in which heat transfers can be reduced.

Reducing heat transfers is important for homes and other buildings, but it is important for other things as well.

A Vacuum flasks reduce the heat transferred from hot drinks to the air, and clothes reduce the heat transferred from your body to the air.

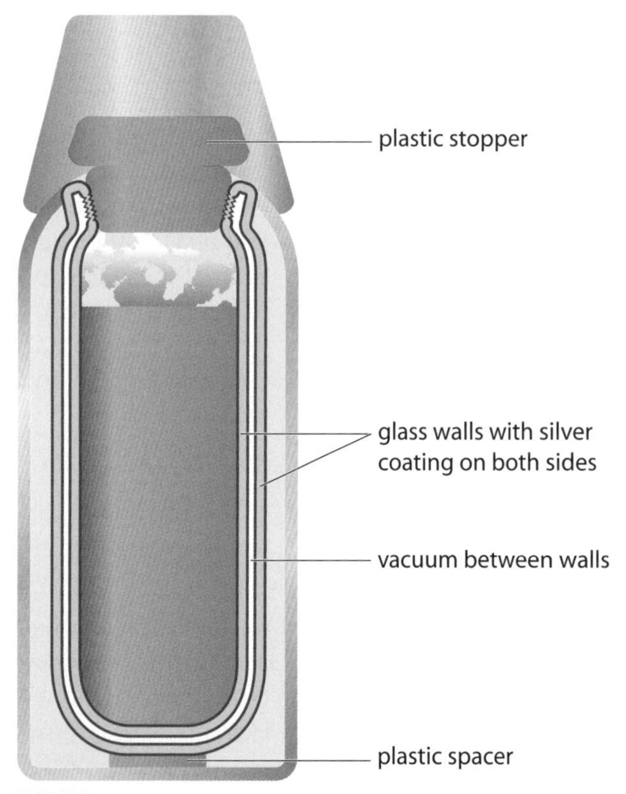

B Vacuum flask.

1 Look at diagram B. List the things that heat energy has to pass through as it is transferred from the coffee to the outside air.

2 Which features of the flask reduce heat transfer by:
- **a** radiation
- **b** conduction
- **c** convection?

3 The flask in photograph A has steel walls instead of glass.
- **a** How would this affect how fast the coffee cools? Explain your answer.
- **b** Why do you think steel is used instead of glass?

4 Would the flask keep a cold drink cold on a hot day? Explain your answer.

Vacuum flasks are made to reduce heat transfer as much as possible. However, some things need to reduce heat transfer in some places and increase it in others.

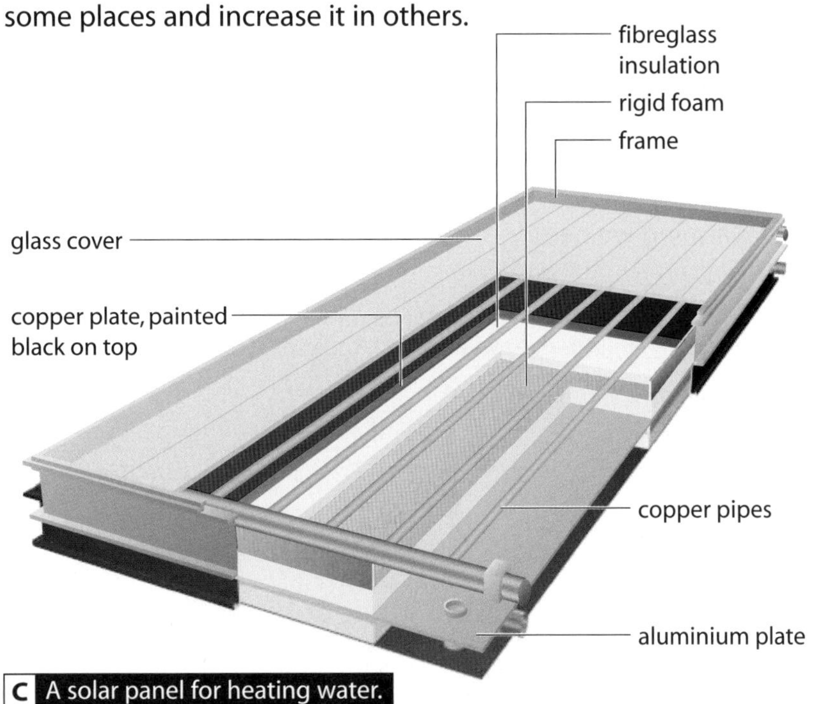

C A solar panel for heating water.

Hot drinks are often sold in paper or polystyrene cups. Sometimes the cups have lids.

D A selection of disposable drinks cups.

7 Many drinks cups are made from insulating materials. Suggest two different reasons for this.

8 A company wants to improve the insulation of their hot drinks cups. They could cover the cup with metal foil, or they could use a lid on the cup.
a How would the foil help to keep the drink in the cup hot?
b How would the lid help to keep the drink hot?
c The company cannot afford to make both changes. Plan an investigation to find out which change is the best at keeping a drink hot.

5 Look at diagram C. Which parts of the solar panel are designed to:
a let infra red radiation go through it
b absorb infra red radiation
c reflect infra red radiation?

6 Which parts are designed to reduce heat transfer by:
a convection
b conduction
c radiation?

9 Draw up a table to show the different ways in which heat transfers can be reduced. You can use ideas from earlier topics as well as from this page.

P D P P

P1a.7

Reducing energy consumption

By the end of this topic you should be able to:

- evaluate the effectiveness and cost-effectiveness of different ways of reducing energy consumption in homes.

If we are going to limit the effects of global warming we need to reduce the amount of energy we use.

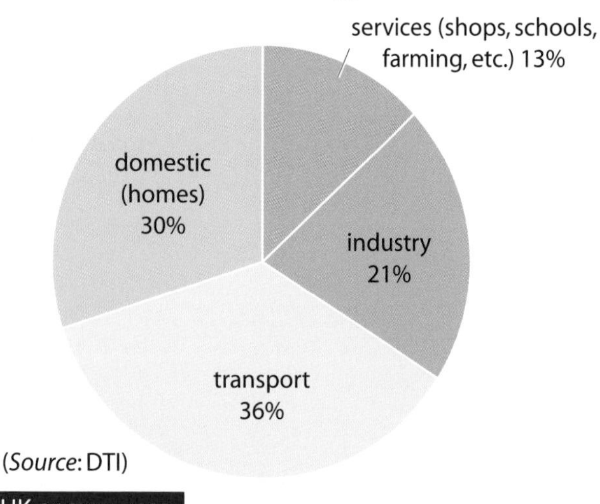

A UK energy uses.

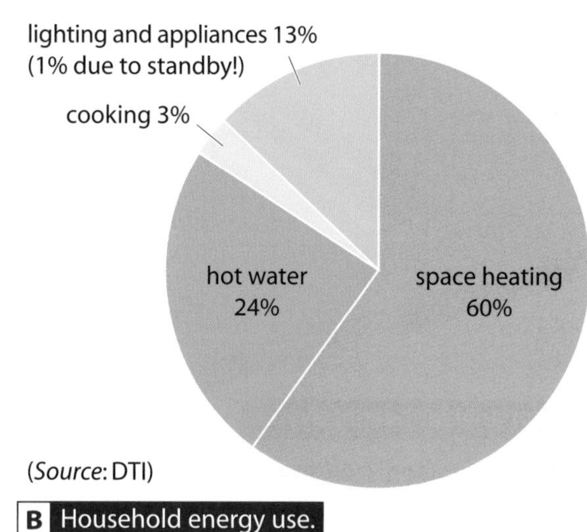

B Household energy use.

If we can reduce the amount of energy we use in our homes, we could help to reduce the amount of carbon dioxide being put into the atmosphere. We cannot all live in houses like the ones in Hockerton, but we can make sure our homes do not use too much energy for **space heating** (heating rooms) or heating water. This will also help to save money on gas and electricity bills. A family living in a three-bedroom semi-detached house could spend up to £500 a year on heating.

1 Look at chart A. Where is most of the energy in the UK used?

2 Look at chart B. Why is it important to try to use less energy for heating?

C What can the owners of these houses do to cut their fuel bills?

There are various ways that people can reduce their energy bills, but most of them cost money. Diagram D shows some of the options. You need to decide the **pay-back time** before you can decide which methods to use. The most **cost-effective** method is the one with the shortest pay-back time.

3 Look at diagram D.
   a Which method is the cheapest?
   b Which method will save the most energy? Explain your answer.

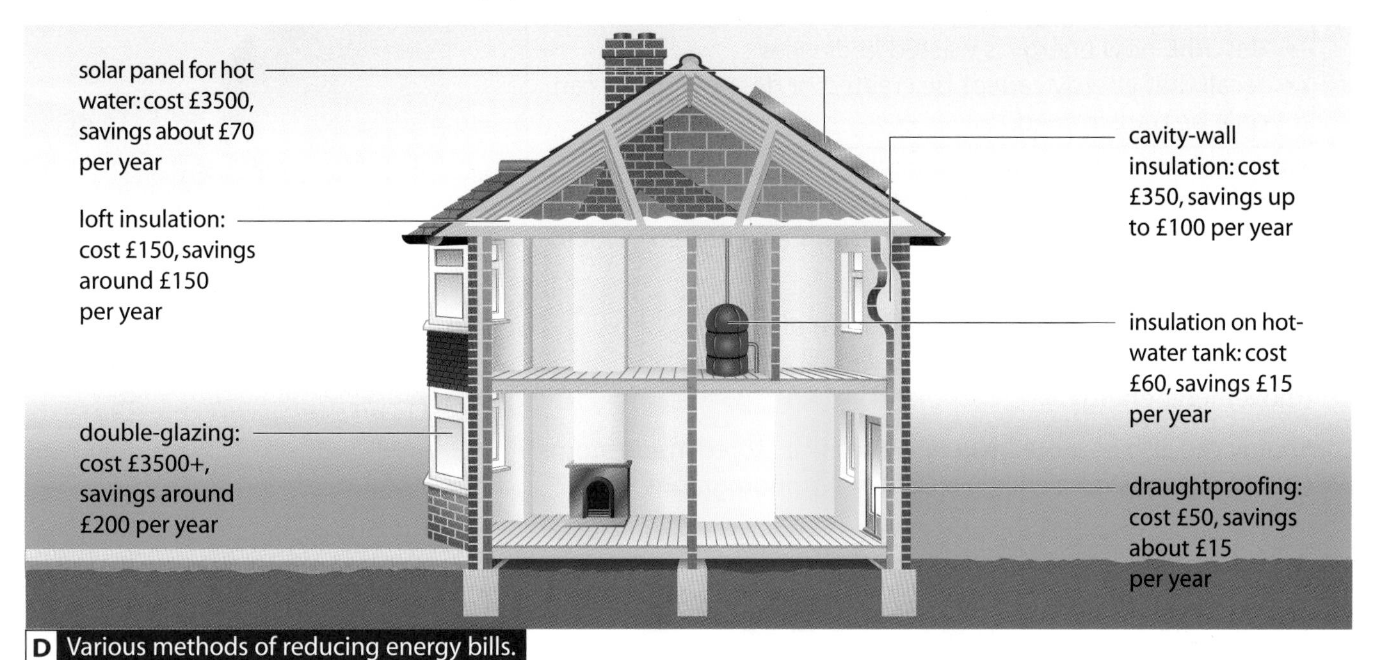

D Various methods of reducing energy bills.

Example

It costs £60 to insulate the hot-water tank, and this will save you £15 a year on energy bills. How long will it take you to save the money on energy bills that it cost you to install the insulation?

$$\text{Pay-back time} = \frac{\text{cost}}{\text{savings per year}} = \frac{£60}{£15} = 4 \text{ years.}$$

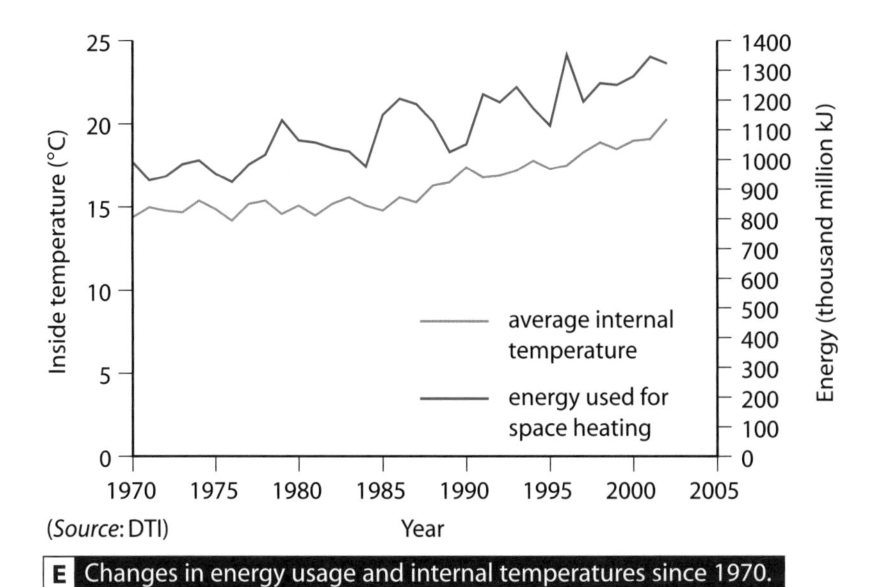

E Changes in energy usage and internal temperatures since 1970.

4 **a** Work out the pay-back times for all the energy-saving methods shown in diagram D.
   b Which method is the most cost-effective?

5 Look at graph E. How could we reduce our energy bills without spending any money? Explain your answer.

6 Why might someone decide to insulate their hot-water tank first, even though it is not the most cost-effective method?

7 Many people can get grants to help with the cost of installing insulation. Calculate the pay-back time for:
   a solar panels, if you get a £400 grant
   b cavity-wall insulation, if you get a £200 grant.

8 Describe the different ways in which energy consumption in a house can be reduced, and how to work out the most cost-effective method.

Energy transformations

By the end of this topic you should be able to:

- describe the energy transformations in a range of devices
- describe how energy is wasted in devices
- recall that energy cannot be created or destroyed, but can only be transformed.

Energy is needed for everything. Without energy we would not exist – there would be no light or warmth, and nothing would happen.

Heat (or thermal) energy is just one form of energy. Other forms of energy are light, sound, electrical and **kinetic** (movement) energy.

Energy can also be stored. **Nuclear energy** is stored inside atoms. Other forms of stored energy are shown in photographs A to C.

A All these things have stored **chemical energy**.

B The people on this ride have **stored gravitational energy** at the top of the slope.

C Anything that is stretched or squashed is storing **elastic potential energy**.

1 Write down three forms of stored energy.

You have seen that heat energy can be transferred by conduction, convection and radiation. Machines also transfer energy, but most machines also **transform** (change) the energy into a different form of energy. For example:

- a torch transforms stored **chemical energy** into electrical energy and then light energy
- the roller coaster carriages in photograph A will transform gravitational potential energy into kinetic energy as they fall.

2 What are the energy transformations in:
a an MP3 player
b a catapult when it is fired?

Sometimes more than one energy transformation is involved. The engine in a car transforms chemical energy into kinetic energy, which is useful. However, some of the energy in the petrol is converted to heat and sound energy. These forms of energy are not useful, so we call them **wasted energy**.

Energy cannot be created or destroyed; it can only be transformed from one form to another. The total amount of energy does not change. If you could measure the total energy stored in the petrol used by the car, it would be exactly the same as the total heat, sound and kinetic energy produced by the engine.

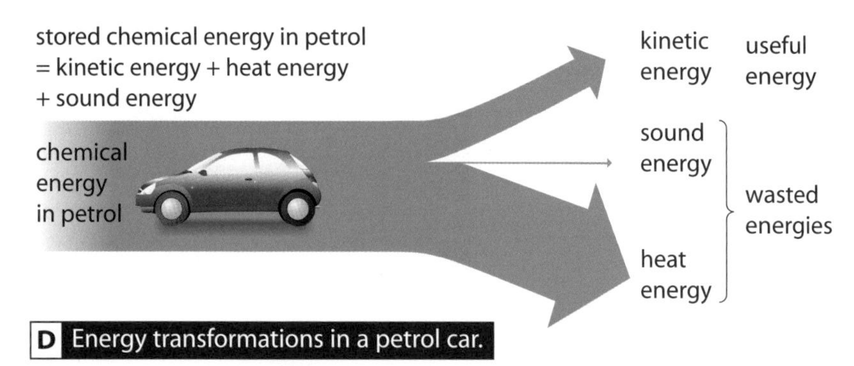

D Energy transformations in a petrol car.

E

3 A light bulb uses 60 J of energy every second and produces 6 J of light energy.
   a What kind of energy does the light bulb use?
   b How much heat energy does it produce every second?
   c Explain how you worked out your answer to part **b**.

4 You are running a race. What forms of energy are you producing that are:
   a useful
   b wasted?

5 Look at the photographs in E.
   a What kind of stored energy is being used in each of the pictures?
   b Write down the useful energy transformation in each picture.
   c What forms of wasted energy are there in each picture?

6 A television transforms 150 J of energy every second.
   a Describe the energy transformations in the television.
   b What is the total energy in these forms of energy?

Electrical energy

By the end of this topic you should be able to:

- give examples of energy transformations in everyday electrical devices
- describe how electricity is transferred from power stations to consumers.

Electricity is a very useful form of energy. It can be transferred long distances in wires, and it can be transformed into lots of other forms of energy. The chemical energy stored in batteries is transformed to electrical energy for use in portable appliances.

1 Look at photograph A. What are the energy transformations that are:
a useful
b wasted?

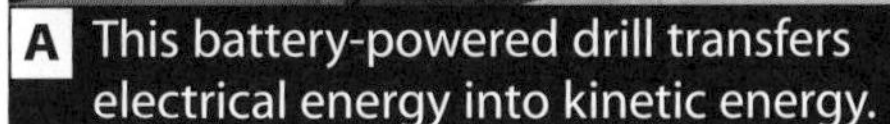
A This battery-powered drill transfers electrical energy into kinetic energy.

B Electricity being used at an outdoor concert.

2 What are the useful energy transformations in photograph B?

3 What are the energy transformations in these electrical devices?
a a kettle
b a television
c a washing machine.
Say which form of energy is useful in each case.

4 Name two electrical devices (other than the ones given on this page) that carry out the following useful energy transformations:
a electrical → kinetic
b electrical → sound
c electrical → light
d electrical → heat.

Most of the electricity we use comes from the **mains supply**. Mains electricity is **generated** in power stations, and is transferred to consumers through the **National Grid**.

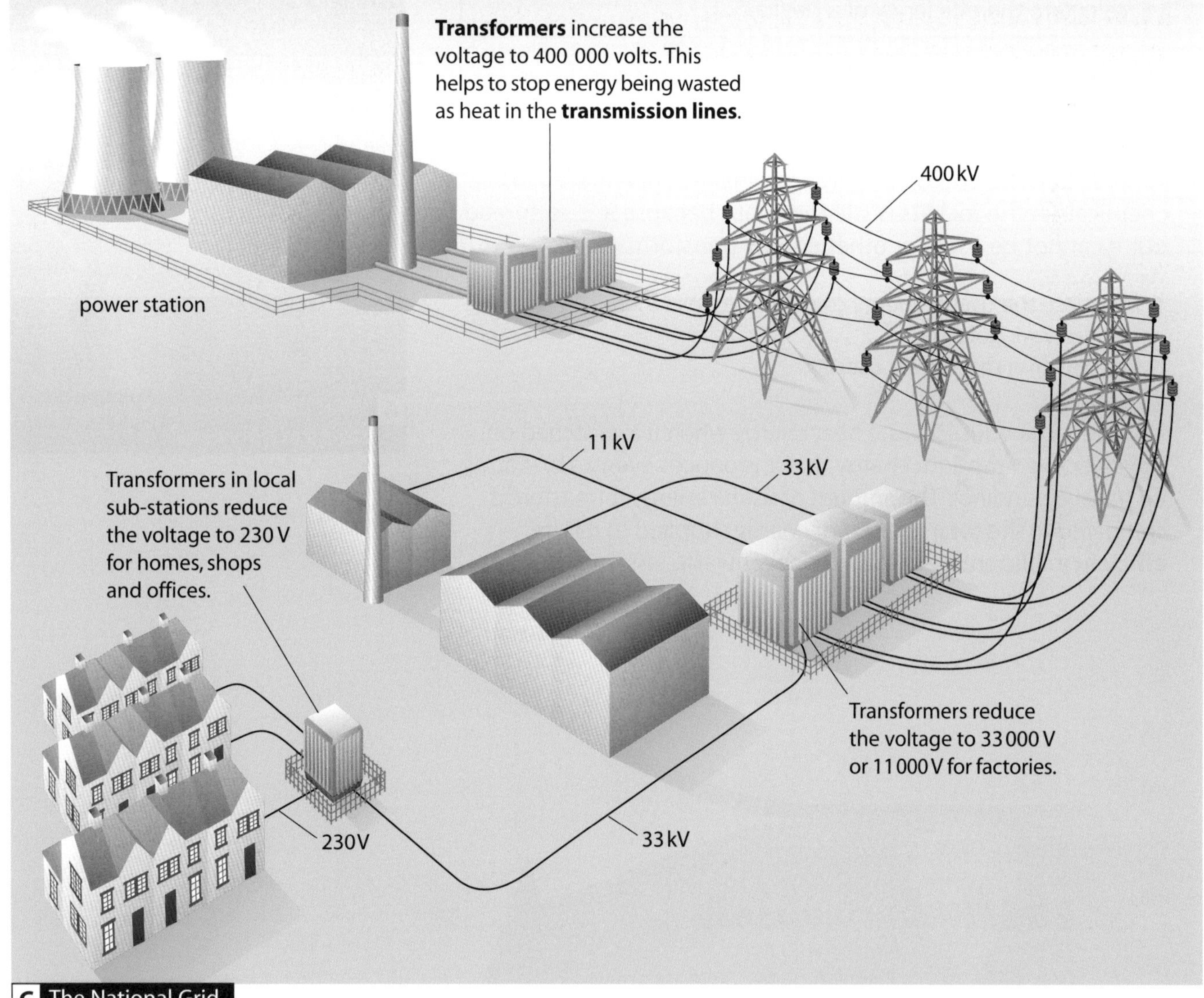

C The National Grid.

5 a What are the advantages of using batteries for the climber in photograph A?
b Why aren't batteries used for the devices in photograph B?

6 a Why is electricity such a useful form of energy?
b How does mains electricity get to our homes?

P1a.10

Efficiency

By the end of this topic you should be able to:

- explain that when energy is transformed, only part of it is transformed into a useful form of energy and the rest is wasted
- explain that all energy is eventually transferred to the surroundings, becoming more spread out and more difficult to use in further transformations
- explain what efficiency means
- calculate the efficiency of a device.

A What happens to the chemical energy stored in a firework?

These fireworks convert stored chemical energy (a form of **kinetic energy**) into light energy, sound energy and heat energy. All this energy is transferred to the surroundings and spreads out. The energy that was originally concentrated in the chemicals in the rockets is still there, but because it is so spread out it cannot be used for other energy transformations.

1 What are the useful energy transformations when:
a a firework goes off
b a washing machine is used?

A light bulb produces waste heat energy when it is switched on. However, even the light energy that it produces eventually warms up the surroundings. The amount of useful energy it transforms compared to the total amount of energy supplied to it is its **efficiency**. Efficient electrical appliances waste less energy.

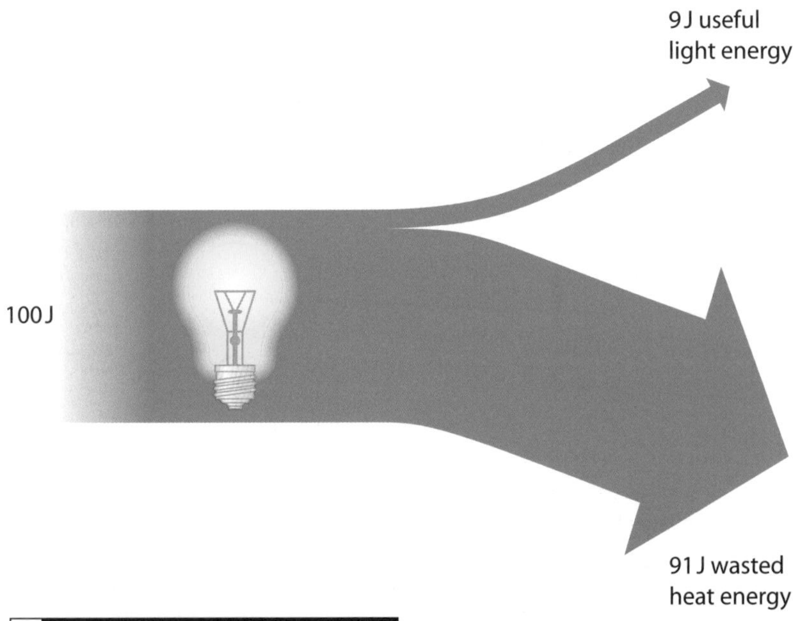

B A light bulb is not very efficient.

The efficiency of an appliance is calculated using this equation:

$$\textbf{efficiency} = \frac{\textbf{energy usefully transferred}}{\textbf{total energy supplied}}$$

The units can be J or kJ, as long as both numbers are in the same units.

Efficiency is sometimes given as a percentage by multiplying the answer by 100%.

Example

What is the efficiency of an electric kettle if it uses 500 kJ of electrical energy and transfers only 400 kJ of energy to the water in the kettle?

$$\text{efficiency} = \frac{\text{energy usefully transferred}}{\text{total energy supplied}} = \frac{400\ \text{kJ}}{500\ \text{kJ}} = 0.8\ (\text{or } 80\%)$$

So the kettle only wastes 20% of the energy that is transferred to it.

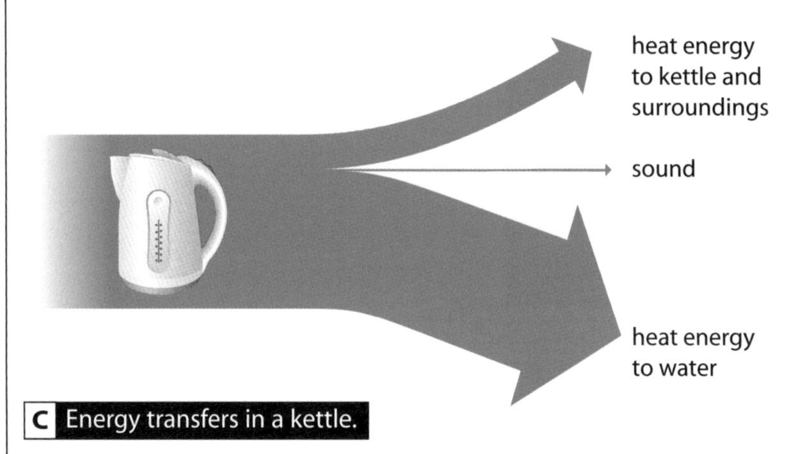

C Energy transfers in a kettle.

If you calculate an efficiency that is bigger than 1 (or bigger than 100%) you have probably put the numbers into the equation the wrong way round!

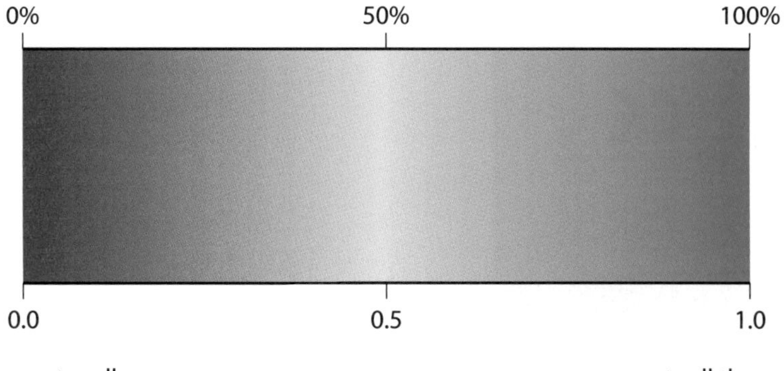

D What efficiency values mean.

2 Look at diagram C. How could you make the kettle more efficient?

3 An electric immersion heater uses 50 kJ of electrical energy and transfers 45 kJ of heat to the water.
  a How efficient is it?
  b What forms of energy are wasted?

4 **a** How efficient is the light bulb in diagram B?
  b An 'energy-saving' bulb gives out the same amount of light, but only uses 18 J of energy per second. How efficient is this bulb?

5 Your muscles waste about 75 J of energy for every 25 J they convert into movement.
  a How efficient are your muscles?
  b Draw a diagram like the one in C to show the efficiency of your muscles.

E An energy-saving bulb (left) and a 'normal' bulb (right).

6 **a** If energy cannot be destroyed, why can't we continue to use it in machines?
  b Explain what efficiency means and how to calculate it.

P D P P

Paying for electricity

By the end of this topic you should be able to:

- calculate the amount of electrical energy used by an appliance
- calculate the cost of this electricity.

We use electricity for a lot of different things, but it all has to be paid for. Electricity bills are worked out from the amount of energy used.

A Powerful speakers transfer a lot of sound energy every second.

The rate at which energy is transferred to an electrical appliance is called **power**, which is measured in **watts (W)**. One watt is one joule of energy being transferred every second. Since many electrical appliances use a lot of power, we usually use **kilowatts (kW)**. 1 kW = 1000 W.

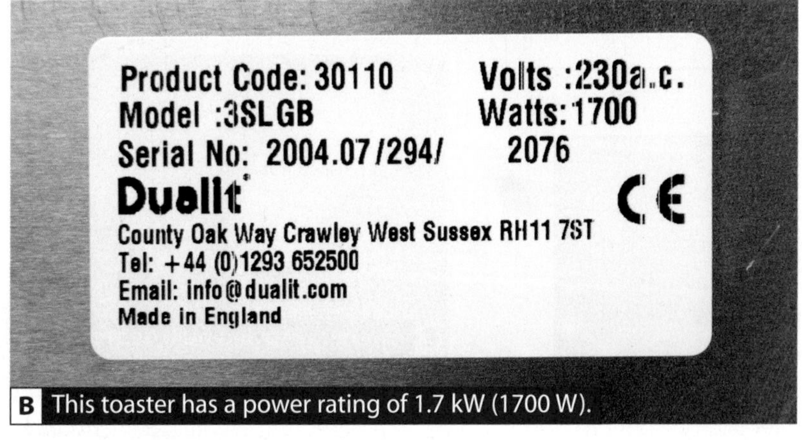

B This toaster has a power rating of 1.7 kW (1700 W).

A joule is quite a small amount of energy. Electricity companies use a much larger unit for electrical energy called the **kilowatt-hour (kWh)**. This is the amount of energy that is transferred by a 1 kW device in 1 hour. It is sometimes called a **Unit** of electricity.

1 **a** What does 'power' mean?
b Write down two different units for measuring power.

2 What is a Unit of electricity?

The energy used can be calculated using this equation:

energy transferred	=	**power**	×	**time**
(in **kilowatt-hours, kWh**)		(in **kilowatts, kW**)		(in **hours, h**)

Example

A 10 kW electric shower is used for a total of 1.5 hours during a week. How many kilowatt-hours of energy does it transfer?

energy = power × time = 10 kW × 1.5 h = 15 kWh

C An electric shower.

Electricity companies usually give their prices in pence per kWh. You can work out the cost of electricity by multiplying the energy by the cost per kWh.

cost	=	**energy used**	×	**cost per Unit**
(**in pence, p**)		(in **kWh** or **Units**)		(**in pence, p**)

Example

An electric heater has a power rating of 2 kW. What is the cost of using the fire for 3 hours if one kWh of electricity costs 5p?

Energy in kWh = 2 kW × 3 h = 6 kWh

Cost in pence = 6 kWh × 5p = 30p

3 How much energy is transferred to a 3 kW electrical heater that is turned on for 5 hours?

4 How much energy is transferred to a 100 W bulb left on for 7 hours?

5 Look at photograph D. An electric lawnmower has a power rating of 1.2 kW.
- **a** It takes 30 minutes (0.5 h) to mow a lawn. How many kWh of electricity does it use?
- **b** If a kWh costs 6p, how much does it cost to mow the lawn?

6 Each bar of an electric fire uses 1 kW of electricity. If electricity costs 5p per kWh how much does it cost to use:
- **a** one bar of the fire for 3 hours
- **b** two bars of the fire for 5 hours?

7 The kWh is a large unit of energy. What is a kWh in joules?

D How much does it cost to cut the grass?

8 Explain how to calculate the energy used by an appliance, and how much this electricity would cost.

P1a.12

Reducing electricity use

By the end of this topic you should be able to:

- compare the advantages and disadvantages of using different electrical devices
- evaluate the effectiveness and cost effectiveness of methods used to reduce the consumption of electrical energy.

About 20% of the energy we use in our homes comes from electricity. Some of this energy is used for heating and cooking, but most of it is used for appliances that can only use electricity.

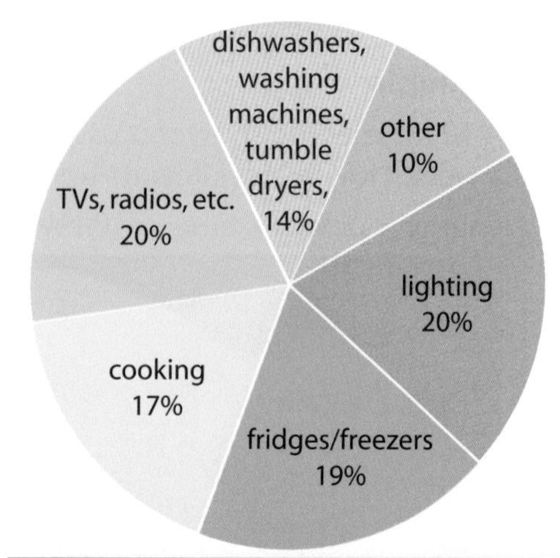

A How electricity is used. This does not include electricity used for space heating.

1 Look at the uses of electricity on chart A.
  a Which appliances could use other sources of energy?
  b Explain your answer.

The amount of energy used by an appliance depends on its power rating and how long it is in use. We can try to reduce the amount of energy we use by reducing either or both of these factors.

Modern appliances are much more efficient than they used to be. A 10-year-old fridge will use about 85 kWh more electricity per year than a new one. However, not all new appliances have the same efficiency. One brand of fridge may be more efficient than another, and often the more efficient models are a bit more expensive to buy.

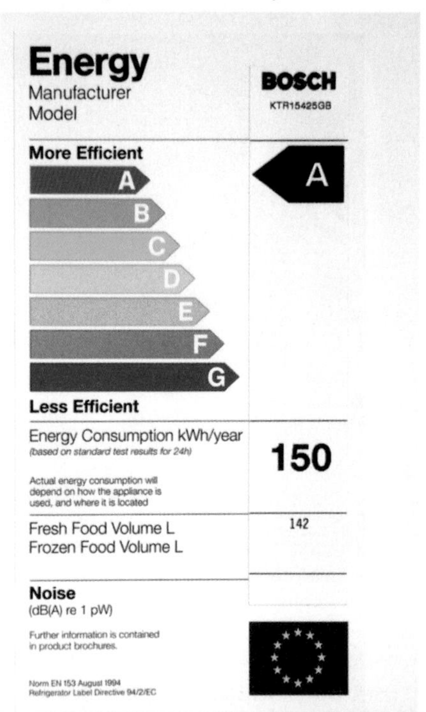

B New appliances have labels to show how efficient they are.

2 If someone replaced their 10-year-old fridge with a new one, how much money would they save each year if electricity costs 6 p per unit?

3 Why might someone buy a more efficient fridge even if it was more expensive?

Sometimes you can use a different kind of appliance to do the same job. For example, microwave ovens heat food in a completely different way to normal ovens, which means the food cooks much more quickly. Because the microwave oven itself does not get hot, it also transfers less wasted heat energy to the surroundings. A microwave oven has an efficiency of nearly 0.6, but a conventional oven has an efficiency of only about 0.15.

JACKET POTATOES

You will need:

1 large potato

Cook for:

$1\frac{1}{2}$–2 hours in a medium oven

15–20 minutes in 850 W microwave oven

C Cooking times for baked potatoes.

4 a Give two reasons why a microwave oven would use much less energy than a conventional oven to cook a piece of food.

b Suggest why people still use conventional ovens for cooking. Give as many reasons as you can.

D The cake on the left was cooked in a conventional oven, the one on the right was cooked in a microwave oven.

The people who live in the houses at the Hockerton Housing Project monitor the energy they use. One very easy way of reducing electricity consumption is by switching things off when they are not in use.

5 Look at the appliances in chart A.

a Which appliances do you think are *really* necessary? Explain your choices.

b Why do you think that people choose to buy appliances that are not really necessary? Give as many possible reasons as you can.

6 Write down all the things you should think about when you are buying a new electrical appliance.

P1a.13

Generating electricity

By the end of this topic you should be able to:

- describe how electricity is generated in most power stations
- recall the common energy sources used in power stations.

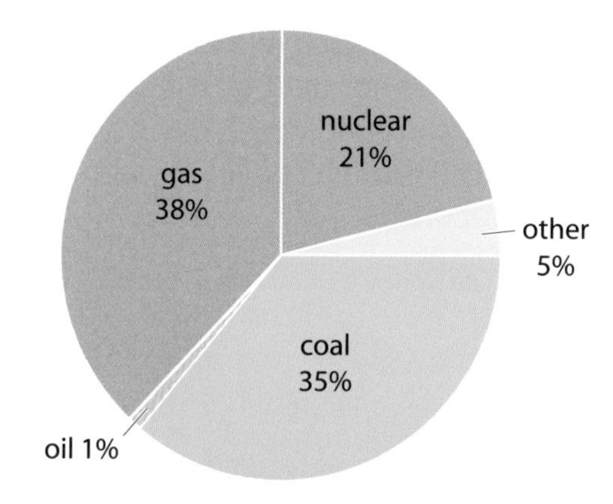

A Energy sources for generating electricity in the UK in 2003.

Electricity is not a source of energy because it has to be generated using other forms of energy. Most of the electricity we use in the UK is generated using **fossil fuels** (coal, oil or natural gas) or nuclear energy.

1 Look at chart A. List the fuels used for generating electricity in order, starting with the one which is used the most.

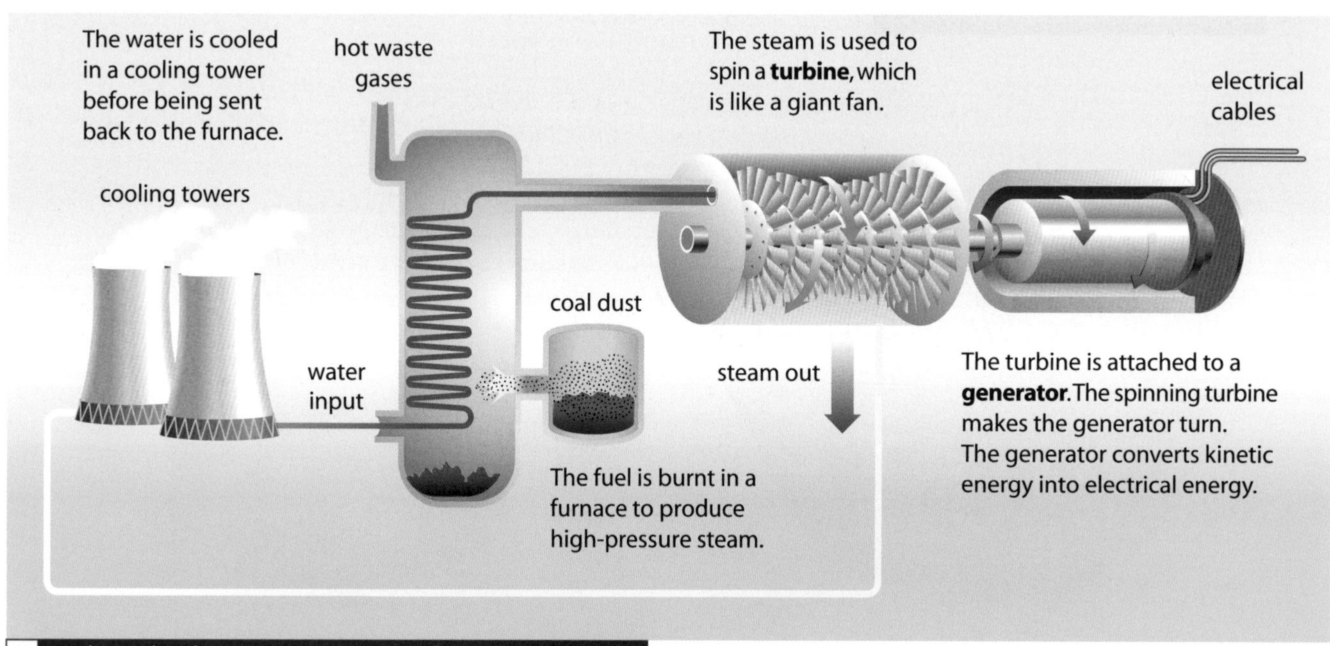

B Look at what happens inside a coal-fired power station.

C A turbine from a power station.

2 What is the National Grid?

3 Why isn't electricity a source of energy?

Fossil fuels are being formed very slowly today underground somewhere on the Earth. However it takes millions of years for these fuels to form, and we are using them up much faster than they are being formed. These fuels are called **non-renewable resources**, because once we have used up the ones that exist at the moment there will be no more left.

Nuclear power stations generate electricity in a similar way to a fossil-fuelled power station, but they use **uranium** or **plutonium** as fuel. These elements are **radioactive**. They do not burn in a chemical reaction like fossil fuels do. Instead, the atoms themselves split up to make new elements. This kind of reaction is called **nuclear fission**. Fission produces a lot of heat from a small amount of fuel, so a nuclear power station has lower fuel costs than a fossil-fuelled one.

Fuel	Estimated time left
Coal	400 years
Oil	60 years
Natural Gas	40 years

D How long our reserves of fossil fuels will last at the rate we are currently using them.

4 Why are fossil fuels called non-renewable resources?

5 Look at table D. Give two reasons why is it difficult to estimate how long fossil fuel reserves will last.

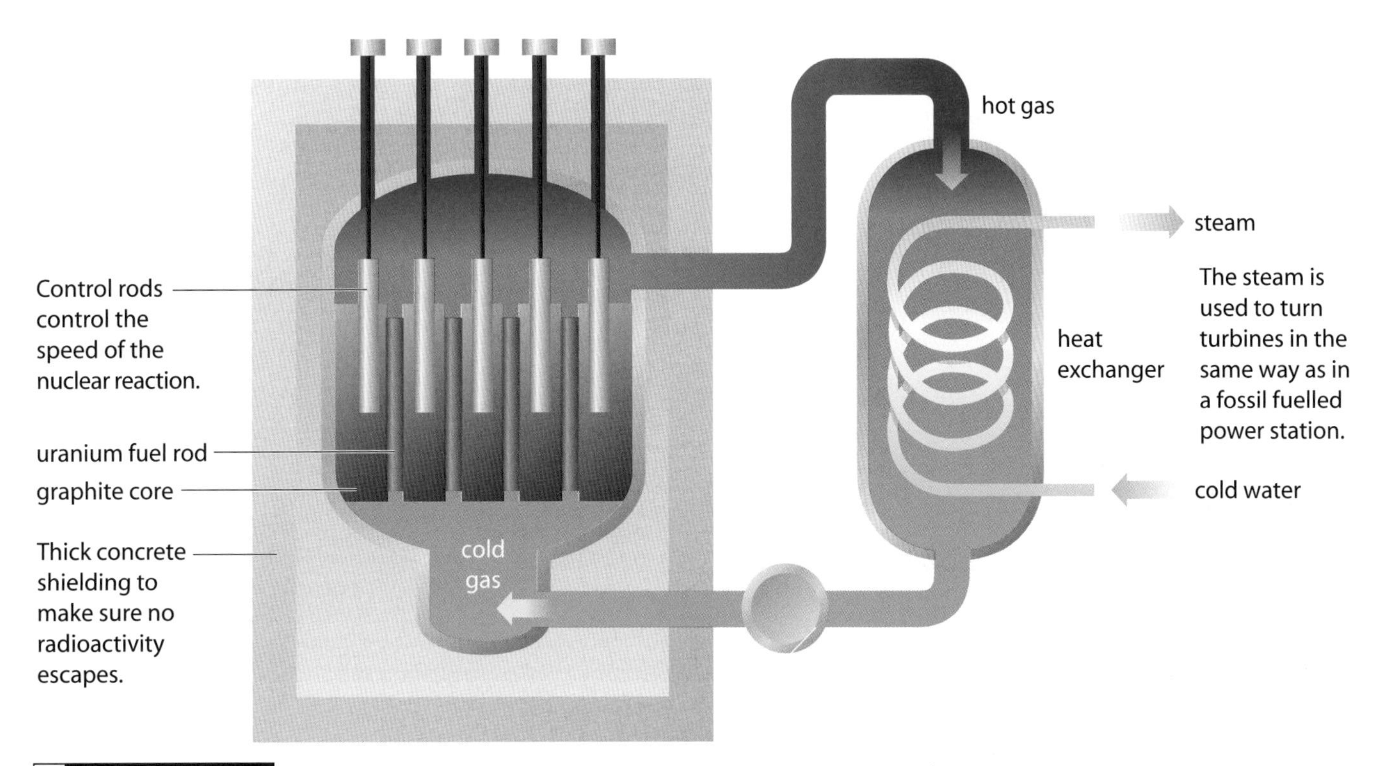

E A nuclear reactor.

6 Name two nuclear fuels.

7 Describe two differences between a fossil-fuelled and a nuclear power station.

8 Uranium is found in the Earth. Is uranium a renewable or a non-renewable resource?

9 Draw a flowchart to show all the stages and energy changes in a power station. You could start your diagram like this:

chemical energy in fossil fuel —burns in furnace→

F

P1a.14

Comparing power stations

By the end of this topic you should be able to:

- describe the effects of fossil-fuelled and nuclear power stations on the environment
- compare the costs involved in different kinds of power stations.

Drax Power Station is the largest coal-fired power station in the UK. It produces 7% of the electricity we use in the UK. Burning coal produces waste gases and also solid waste.

A Drax power station.

Waste	What happens to it?	What can be done?
Carbon dioxide	goes into the atmosphere; contributes to global warming	capture it and store it so it does not go into the atmosphere
Sulphur dioxide	goes into the atmosphere; causes acid rain	remove it from the waste gases, or from the coal before it is burnt
Nitrogen oxides	goes into the atmosphere; causes acid rain	use furnaces that reduce the amount produced
Dust	smoke and dust gets into the atmosphere	remove from the waste gases
Ash	most is sold for building; some goes to landfill sites	cover landfill sites with soil and grow trees or crops on them

B

1 List the ways in which coal-fired power stations can cause pollution.

Natural gas is a cleaner and more efficient fuel as it does not contain sulphur and has less nitrogen than coal. Power stations that burn natural gas emit less carbon dioxide for each Unit of electricity they produce.

Nuclear power stations do not emit any carbon dioxide or other gases. However, there are other potential problems with nuclear power stations. The waste they produce is radioactive, and some of it will stay radioactive for millions of years. The waste must be sealed into concrete or glass and buried safely so the radioactivity cannot damage the environment. A nuclear power station also needs to be carefully **decommissioned** (dismantled safely) at the end of its life so that no radioactivity escapes into the environment. It costs a lot more to build and to decommission a nuclear power station than a fossil-fuelled one.

There are not many accidents in nuclear power stations, and the power stations are designed to contain any radioactive leaks. However, if a major accident occurs it can have very serious consequences.

The world's worst nuclear accident was at Chernobyl, in the Ukraine, in 1986. It killed 30 people immediately, and many people became ill from the effects of the radiation. Altogether, over 300 000 people had to leave their homes permanently because the area around the nuclear plant was contaminated. Winds blew radioactive dust across many other countries, including the UK.

2 Describe two advantages of using natural gas instead of coal to generate electricity.

3 a Write down one advantage of nuclear power over fossil-fuelled power stations.
  b Write down two disadvantages.

4 a Why do you think a nuclear power station costs more to build?
  b Why is decommissioning it properly very important?

C Sheep from parts of Cumbria, Wales and Scotland still have to be tested for radiation before they can be sold.

D The reactors at Chernobyl.

5 How can a nuclear accident affect people in other parts of the world?

6 If a new power station was going to be built near your town, which fuel do you think it should use? Explain your answer.

7 Make a table to show the advantages and disadvantages of coal, gas and nuclear power stations.

Renewable resources

By the end of this topic you should be able to:

- recall the renewable energy resources that can be used to drive turbines directly
- explain how solar cells and geothermal energy can be used to generate electricity.

Fossil fuels are referred to as non-renewable resources because they will run out sometime in the future. However, there are other sources of energy that we can use to generate electricity that will not run out. These are called **renewable** resources. At present, almost all of the electricity we use comes from non-renewable resources.

Many renewable energy resources are used to turn turbines which then turn generators. For example, dams can be built in hilly areas to trap water in reservoirs. The water is made to flow downhill in pipes to a **hydroelectric** power station, where the kinetic energy in the water is used to generate electricity.

B These **wind turbines** off the coast of Norfolk can each generate up to 2 MW of electricity.

The tides can also be used to generate electricity. One way of doing this is to build a huge **barrage** across a river estuary. Turbines fitted in the bottom of the barrage are turned as the tides flow in and out. The tides also cause currents in the open sea, and turbines can be mounted on the sea bed to use this source of energy. This technology is still being developed.

A The solar cells in these panels generate over 5500 kWh per year. They cost just over £40 000 when they were installed in 2002.

1 Why is it important to use renewable resources for generating electricity?

2 How can energy be produced using:
a the Sun
b wind?

C Waves force air up and down a tube inside the concrete building. The moving air turns a turbine.

D A possible design for a tidal-stream turbine.

3 Describe four different ways in which water can be used to generate electricity.

Hot rocks can also be used to generate electricity. In some volcanic areas hot water and steam rise to the surface, or pipes can be drilled into the ground to allow the steam to rise. This steam can be used to drive turbines, which drive generators. This is known as **geothermal** energy. This is not renewable in the same sense as the other sources on these pages, as eventually the hot rocks will cool down, but the supply of heat will last much longer than supplies of fossil fuels.

E A geothermal power station in Iceland. Electricity cannot be generated economically in this way in the UK.

Although there are no fuel costs for any of these renewable energy resources, money has to be spent to make solar panels, or to build turbines or dams. So although the energy itself is free, it costs money to turn the energy into electricity.

4 **a** How does a geothermal power station work?
b How is this different to a fossil-fuelled power station?
c How is it similar to a fossil-fuelled power station?
d Why can't this source of energy be used in the UK?

5 Look at all the renewable resources on this page. Which ones can generate electricity:
a all the time
b only some of the time.

6 Draw up a table to list all the different sources of renewable energy and briefly describe how each one works.

Renewables and the environment

By the end of this topic you should be able to:

- describe the effects that different renewable energy resources have on the environment.

People often think of renewable energy resources as 'clean' sources of energy, because most of them do not add polluting gases to the atmosphere. However, they can affect the environment in other ways.

A Wind turbines close to a house in Cold Northcott, Devon.

One objection that many people have to wind turbines is that they cause **visual pollution** (they spoil the view). Fossil-fuelled power stations are not pretty, but people have got used to them, and they are usually built in areas that already have industrial buildings. **Wind farms** are often built in places where people go to enjoy the countryside. Some people also complain about the noise they make, and say that they take up too much land in the countryside.

1 **a** Why are people more concerned about the visual impact of wind turbines than power stations?
b Some wind farms are built in the sea. What problems could there be with this?

2 What effects might wind turbines on land have on:
a wildlife
b farming?

Hydroelectric power stations need large reservoirs to supply the falling water. In most cases these reservoirs are created by damming a valley and flooding it. This destroys habitats in the valley, and can actually cause carbon dioxide and methane to be added to the atmosphere for a while as the flooded plants die and rot.

B These pipes are used to supply water to generate electricity in a hydroelectric power station at the bottom of the hill. They can be fed through tunnels in the hill to reduce the visual impact.

Tidal barrages are huge dams that change the flow of rivers. This may affect birds and other wildlife that live or feed on tidal mudflats, and may affect the migration of fish.

C These birds feed off animals that live in the mud. These may not survive if the flow of water in the river is changed.

Waste gases

The main problem with fossil-fuelled power stations is that they emit carbon dioxide into the atmosphere. In geothermal power stations, the hot water rising from the depths of the Earth often has some gases dissolved it. These gases can escape into the atmosphere and include carbon dioxide and methane, which contribute to global warming. However, a geothermal power station only produces about one-tenth of the amount of carbon dioxide as a coal-fired power station for the same amount of electricity generated.

3 How could a hydroelectric power station cause atmospheric pollution?

4 **a** Which parts of a hydroelectric scheme could cause visual pollution?

b How could this be kept as low as possible?

P

5 List the effects of a tidal barrage on:

a scenery

b shipping

c wildlife.

6 What might a person who is keen on water sports think about:

a a tidal barrage

b a reservoir for hydroelectricity?

P

7 Suggest how a wave power generator might affect the environment (photograph C on page 197).

8 Draw up a table to summarise the environmental effects of each source of renewable energy.

D

P

P

Meeting the demand

By the end of this topic you should be able to:

- describe how the demand for electricity changes
- recall that different power stations have different start-up times
- recall that electricity from some renewable resources is not available continuously.

The demand for electricity changes during the day and also during the year.

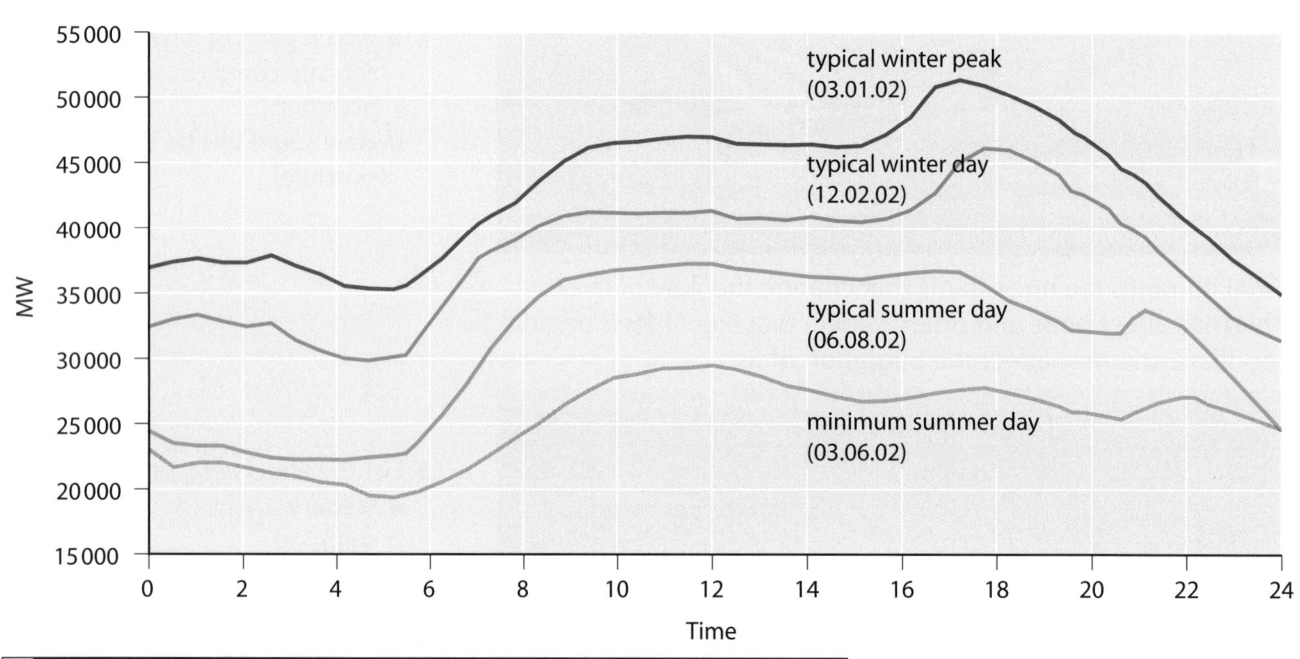

A Comparison of electricity demand through the day in summer and winter.

1 Look at graph A.
  a Describe the demand for electricity on a typical winter day.
  b Explain why the demand changes during the day.

2 a Why is there a peak in demand in the early evening in winter but not in summer?
  b Why is there more demand for electricity in winter than in summer?

The demand also changes from minute to minute. If there is a popular programme on TV, millions of people sit down to watch it and the overall demand for electricity falls. At half-time in a football match, or when the adverts come on, thousands (or even millions) of people get up and switch on lights or kettles, and the demand can shoot up by over 2 MW. This sudden change is called a **TV pick up**.

B The biggest pick up experienced by the National Grid was just after the total solar eclipse in August 1999.

Power engineers need to predict these changes because it is not easy to suddenly increase the amount of energy generated. It can take hours to start up a coal-fired power station from cold, so power stations are often kept running below their full generating capacity. This can waste some energy, but it means that the amount of electricity generated can be increased within minutes instead of hours. Hydroelectric power stations are very useful because they can start producing their maximum amount of electricity within a minute.

Supplying enough electricity will be even more complicated when more renewable resources are used. Many sources of renewable energy are only available some of the time, and we cannot always predict when they will be available.

Power stations run more efficiently when they are generating their maximum amount of electricity, so sometimes more electricity is being generated than is needed. Some of this energy can be stored using a **pumped-storage** power station.

3 a Why do power engineers need to know what weather is forecast for the next few hours?

b Why do they sometimes need to know what the TV schedules are?

4 Why can't power engineers just use hydroelectricity to meet sudden demands? (*Hint*: you may need to look back at Topic P1a.13.)

5 Suggest a circumstance in which hydroelectricity might not be available.

6 Other than hydroelectricity, which resource is available at predictable times?

7 When might wind power be available but not solar power?

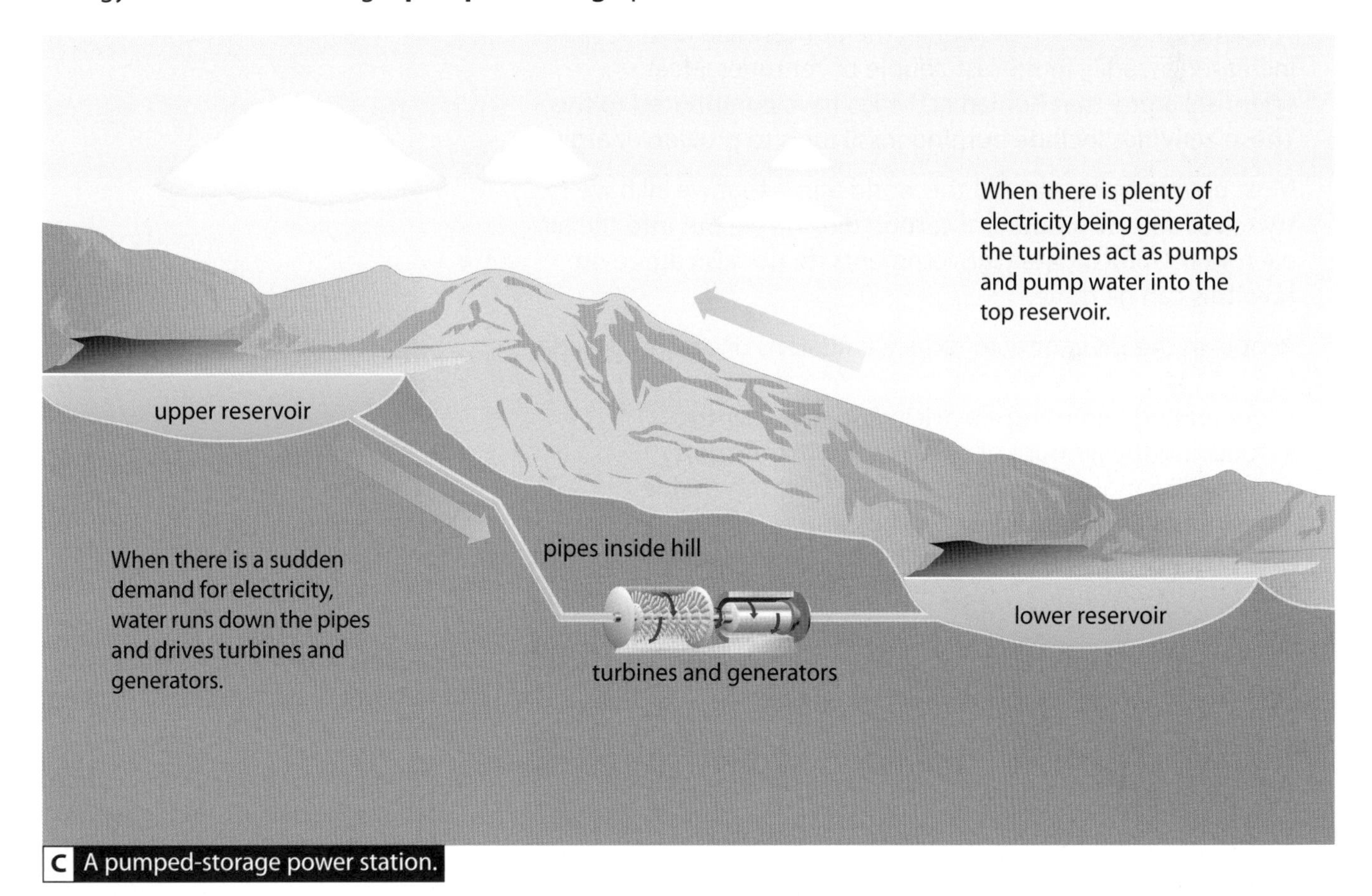

C A pumped-storage power station.

8 Look at graph A. At what times of day would a pumped-storage power station be:

a pumping water up to the top reservoir

b generating electricity?

9 Write a short paragraph explaining why it is important that power engineers can forecast the demand for electricity.

P1a.18

The future

By the end of this topic you should be able to:

- consider some of the advantages and disadvantages of using different energy resources, including the location in which the energy is needed
- understand that a mix of different resources is likely to be the best solution to energy supplies.

Global warming threat to wildlife

MALARIA IN THE UK?

MORE FLOODING LIKELY

A The impact of global warming.

The amount of carbon dioxide in the atmosphere has increased steadily in the last couple of centuries. Most scientists agree that human activities have contributed to this. These activities include burning fossil fuels to provide energy.

Most governments around the world agree that we all have to try to reduce the amount of carbon dioxide we put into the air each year. Unfortunately, governments do not also agree on how this can be done.

People in the UK can try to reduce emissions of carbon dioxide by:

- generating and using electricity more efficiently
- reducing the amount of energy used in transport
- providing more heat and electricity from renewable resources or from nuclear power stations.

B This car is used by the people who live in the Hockerton houses. It runs on electricity from rechargeable batteries, so it produces no emissions.

1 Suggest some ways in which we could reduce the amount of energy used for heating houses.

2 What problems are there in using renewable resources for generating electricity?

3 Why do some people think that we should not be using nuclear energy to generate electricity?

4 Look at photograph B. Is this car really a 'zero-emission' car?

Combined heat and power

A coal-fired power station has an efficiency of only 35%. Gas-fired power stations are more efficient, but still waste a lot of energy as heat. In some places this waste heat energy is used to supply hot water for heating to nearby homes, offices and factories. This is called **combined heat and power (CHP)**.

5 Why can't all homes have their hot water supplied by CHP power stations?

Distributed production

At present, most of the electricity in the UK is generated in large power stations and transmitted for long distances by the National Grid. Offshore wind farms send electricity through undersea cables. Some energy is wasted as heat energy in all these transmission lines.

In the future it may be possible to produce electricity in more places. For instance, CHP could be more widely used if each town had its own small power station. This is called **distributed production**.

C Water flowing through the gates in this weir could be used to generate electricity for local buildings.

D Houses in remote locations could generate most of their own electricity.

6 **a** Where are the best places for generating electricity from the wind?
  b Where in the country is the most electricity used?
  c Explain why a lot of electricity from renewable resources could end up being wasted as heat energy.

7 Suggest two ways in which distributed production could increase the efficiency of the electricity supply.

8 **a** Why do we need to reduce the amount of carbon dioxide we put into the atmosphere?
  b How can this be done?

Assessment exercises

Part A

Multiple choice questions

1 Which of these is NOT likely to increase the demand for electricity.
- **a** A very hot spell in the summer.
- **b** The end of the FA Cup final on the television.
- **c** A spell of mild weather in the winter.
- **d** A spell of very cold weather in the winter.

(*1 mark*)

2 A torch uses 1.5 J of electrical energy from a battery every second, and produces 0.27 J of light energy. What is the efficiency of the torch bulb?
- **a** 5.5
- **b** 0.405
- **c** 0.18
- **d** 0.18 %

(*1 mark*)

3 TVs, computers and radios all waste some of the energy they transform. What form of energy is wasted?
- **a** heat
- **b** light
- **c** sound
- **d** electrical

(*1 mark*)

4 The diagram below shows a coal-fired power station.

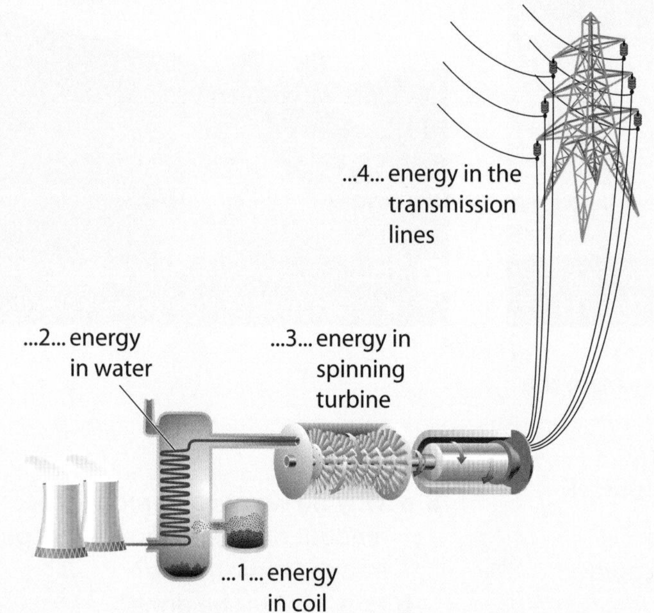

Choose from the different kinds of energy, listed **a–d** below, to complete labels **i–iv**.
- **a** heat
- **b** electrical
- **c** chemical
- **d** kinetic

(*4 marks*)

5 The diagram below shows an insulated box used to keep food hot.

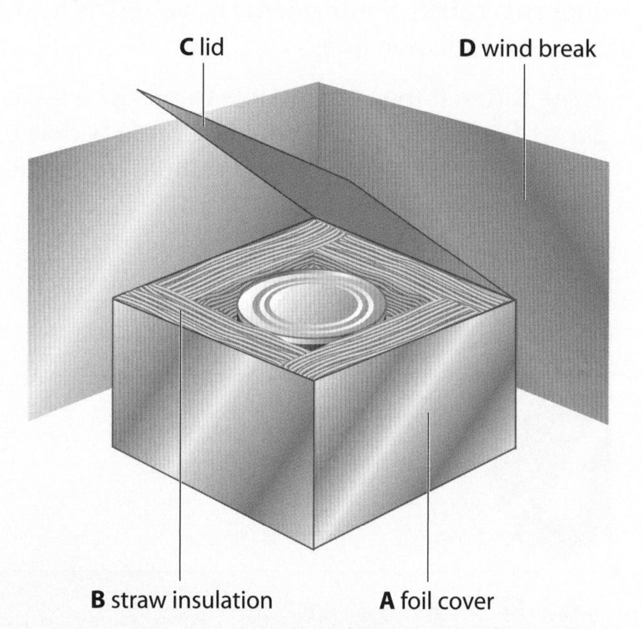

Choose the correct parts, **a–d**, to fill in the gaps in these sentences.
- **i** The ____________ helps to stop the box emitting infrared radiation.
- **ii** The ____________ helps to stop heat being transferred by convection.
- **iii** The ____________ helps to stop heat being transferred by conduction.
- **iv** The ____________ stops the wind moving warmed air away from the box.

(*4 marks*)

6 Asha was testing different ways of insulating a food box. She carried out four different tests, each using materials in different ways. Look at tests **a–d** and decide which kind of variable was being used. Then complete the table accordingly.

a Foam 1 mm thick, 2 mm thick or 3 mm thick.
b Foam, foil or bubble wrap.
c One, two or three layers of foam.
d Light, medium or dark paper.

	Type of variable	Test a, b, c or d?
i	categoric	
ii	ordered	
iii	continuous	
iv	discrete	

(4 marks)

7 Match terms **a–d** with spaces **i–iv** in the paragraph below.

a carbon dioxide
b renewable
c global warming
d fossil

(4 marks)

We need to reduce the amount of **i**______ fuels we burn because **ii**______ being added to the atmosphere is causing **iii**______. We can do this by using more **iv**______ resources to generate electricity.

8 Match terms **a–d** with spaces **i–iv** in the paragraph below.

a thermal conductors
b transparent
c thermal insulator
d electrical conductors

(4 marks)

Materials are chosen for their properties, but sometimes different properties are important. For example, saucepans are made from metal because metals are good **i**______ and they are used for electricity transmission lines because they are good **ii**______. Glass is used for windows because it is **iii**______, and it is used in vacuum flasks because it is a good **iv**______.

Total (Part A) 23 marks

Part B

1 Describe how a convection current will form in the air near an ice-lolly. Draw a diagram to illustrate your answer. *(4 marks)*

2 A TV uses 90 W of power when it is switched on. Electricity costs 6.5 pence per Unit.

a What is 90 W in kW? *(1 mark)*
b How much energy does the TV use in 6 hours? *(1 mark)*
c How much does this electricity cost? *(1 mark)*
d Why should you switch a TV off when it is not being used, instead of putting it on standby? *(1 mark)*

3 **a** Write down two disadvantages of using energy from uranium to generate electricity. *(2 marks)*
b Write down two advantages of using uranium. *(2 marks)*

4 **a** Explain what the terms 'pay-back time' and 'cost-effective' mean. *(2 marks)*
b Describe how you would decide which type of insulation to add to a house. *(2 marks)*

5 Nerys works in a laboratory that tests appliances to see how efficient they are.

a Why are standard procedures necessary in testing laboratories? *(1 mark)*
b Nerys is testing different makes of fridge. What is the dependent variable in her investigation? *(1 mark)*
c Write down two things that Nerys will have to keep the same to make sure her investigation is a fair test. *(2 marks)*

6 **a** What is a TV pick-up? *(1 mark)*
b Why do power engineers need to use weather forecasts? *(1 mark)*
c Which kind of power station can start generating in the shortest time if there is an increase in demand? *(1 mark)*
d Describe one disadvantage of having a large proportion of the country's electricity generated from renewable resources. *(1 mark)*

Total (Part B) 24 marks

Investigative Skills Assessment

Joe investigated the power produced by solar cells with different areas. He used card to block out some of the cells on a solar buggy and then used a speed sensor to measure the speed that the buggy could reach.

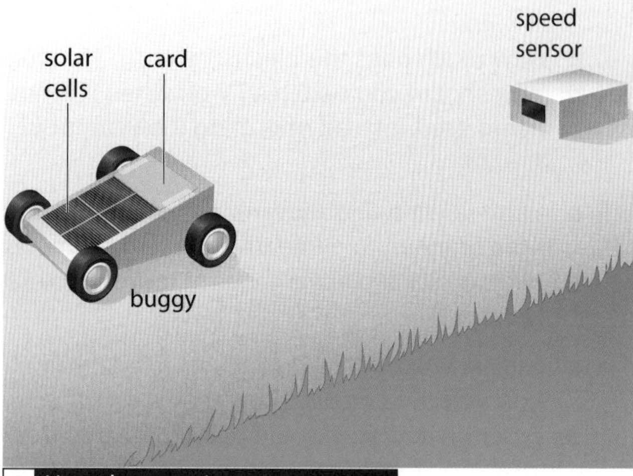

A How the experiment was set up.

Joe's results are shown in Table B, below. He carried out each test three times.

Number of cells left uncovered	Speed of buggy (m/s)		
	1st test	2nd test	3rd test
1	0.29	0.31	0.28
2	0.59	0.58	0.63
3	0.92	0.79	0.87
4	1.11	1.22	1.20
5	1.52	1.48	1.51
6	1.77	1.81	1.79

B Investigation results.

1 a Two of Joe's results do not fit the pattern. Which results are they? (*1 mark*)

b Calculate the average speed of the buggy for each number of solar cells uncovered. Ignore the results that do not fit the pattern that you identified in part a. (*1 mark*)

2 Plot a graph to show Joe's results. (*3 marks*)

3 Look at the results for the test with 1 solar cell uncovered. What is the range of data for this test? (*1 mark*)

4 a What was the independent variable in Joe's test? (*1 mark*)

b Was the independent variable continuous, discrete or categoric? (*1 mark*)

c What was the dependent variable? (*1 mark*)

d Was the dependent variable continuous, discrete or categoric? (*1 mark*)

e Write down one thing that Joe had to keep the same to make his test fair. (*1 mark*)

5 What is the relationship between number of cells and the speed of the buggy shown in Joe's graph? (*1 mark*)

6 Joe was trying to answer this question:
Is the voltage produced by a solar cell proportional to its area? Was Joe trying to measure the voltage directly or indirectly?
Explain your answer. (*2 marks*)

7 ✎ What assumption did Joe make when he decided to answer the question above using a speed sensor? Suggest a better way of investigating the question. (*4 marks*)

8 Sally repeated the test the next day. The average speed of her buggy with all the cells uncovered was 1.85 m/s. Suggest why she might have got an average faster speed than Joe. (*2 marks*)

Total = 20 marks

Glossary

absorb Take in, or soak up.
barrage A dam built across a river estuary.
chemical energy Energy stored in batteries, food and fuels such as petrol.
combined heat and power (CHP) A power station where waste heat is used to provide hot water to nearby houses and other buildings.
conduction The way that heat passes through solids by passing from particle to particle.
convection The way that heat travels through liquids or gases as the particles in them move about.
convection current A flow of liquid or gas caused by part of it being hotter or colder than the rest.
cost-effective A cost-effective change is one where you get back the money you spent on the change quite quickly by savings on energy bills.
decommission To dismantle safely.
dense Something that is heavy for its size.
distributed production The generation of electricity by lots of small power stations or renewable resources spread around the country.
efficiency The fraction of energy that is usefully transferred.
elastic potential energy Energy stored in something that is stretched or squashed.
electromagnetic wave Waves such as light, infra red and radio waves.
emit Give out.
fluid A liquid or a gas.
fossil fuel A fuel formed from the remains of animals or plants. Coal, oil and natural gas are the main fossil fuels.
geothermal Energy from hot rocks.
generator A machine that makes electricity when it turns.
gravitational potential energy Energy stored in something that is in a high position.
hydroelectric Electricity generated using falling water.
infra red radiation Electromagnetic radiation that we can feel as heat. It has a longer wavelength than visible light, but a shorter wavelength than microwaves.
insulated Covered with insulating material, to stop heat energy transferring in or out.
kilowatt (kW) A unit for measuring power. 1 kW = 1000 W.
kilowatt-hour (kWh) The amount of energy transferred in an hour by a 1 kW appliance.
kinetic energy Energy in moving things.
mains supply The electricity that we get from sockets in the wall.
matt Not shiny.
metal A strong, shiny material that is a good conductor of heat and electricity.
National Grid The system of wires and transformers that distributes electricity around the country.
non-metal Element that is not a metal.
non-renewable resource A resource that cannot be replaced once it has been used. Non-renewable resources will eventually run out.
nuclear energy Energy stored inside atoms.
nuclear fission When atoms of elements such as uranium break up and release energy.
particles Tiny particles of soot, ash and sulfur compounds formed when coal, oil and wood burn. They lead to global dimming.
pay-back time The time it takes to get back (in energy savings) the money you spent on making a change.
plutonium A radioactive element that can be used as a fuel in nuclear power stations.
power How quickly something transfers energy.
properties Ways of describing how a substance behaves.
pumped storage A power station that can pump water to a high reservoir when there is spare electricity being generated, and then use the falling water to generate electricity when more is needed.
radioactive Something that gives out dangerous radiation (not infra red radiation) when it decays.
reflect When something bounces off an object, such as light bouncing off a mirror.
renewable A resource that can be used again and again (such as solar power).
solar panel A black board with water pipes beneath it that is designed to absorb infra red radiation from the Sun.
space heating Heating rooms.
thermal conductors Materials that allow heat to be transferred through them easily.
thermal insulators Materials that allow electricity to flow through them.
thermal radiation Another word for infra red radiation.
transform Change.
transformer A machine that changes the voltage of electricity.
transmission lines The wires hanging from pylons that carry high voltage electricity around the country.
turbine A machine that is turned by a moving fluid.
TV pick up A sudden increase in the demand for electricity caused by people switching on kettles, etc. when a TV programme ends.

Unit A unit for measuring the amount of energy transferred. A Unit is the same as a kilowatt-hour.

uranium A radioactive element that can be used as a fuel in nuclear power stations.

visual pollution Something that looks ugly or spoils the view.

wasted energy Energy that is not useful, such as the heat and sound energy produced by a car engine.

watt (W) The unit for measuring power. 1 watt is 1 joule being transferred every second.

wind farm A lot of wind turbines close together.

wind turbine A kind of windmill that generates electricity using energy from the wind.

Radiation and the Universe

A Making an airport run smoothly is not easy.

B Space science is often the place where new technology is developed. Communications with satellites uses microwaves. Plane navigation is now carried out using satellites.

C Airport staff often use walkie-talkie radios.

D We usually think of radioactivity as a dangerous thing. Radioactivity is not all bad news though. It can be used in smoke detectors and, with jet fuel stores, airports have to have the very best fire alarm systems.

Imagine trying to run Heathrow Airport. Think of all the things people need to do at an airport and how you could manage to get all these thing to work together smoothly. Then remember that you also need to make sure everything happens safely. Think about the security of the airport and the planes.

To get all the passengers onto the aeroplanes without long delays airports use a lot of high-tech equipment. The X-ray machines that scan luggage, computer networks and CCTV cameras have something in common: they all use electromagnetic waves. Light is a type of electromagnetic wave. In this unit you will see how different types of electromagnetic waves are different and how they are similar.

By the end of this unit you should:

- be able to give some uses and hazards of the waves that form the electromagnetic spectrum
- be able to give some uses and dangers of emissions from radioactive substances
- describe what we know about the origins of the Universe and how it continues to change.

1 Write down ten things that happen at an airport.

2 List all the people who might use radios in an airport.

3 List as many uses for radiation that are good for people as you can think of.

Electromagnetic waves

By the end of this topic you should be able to:

- explain how electromagnetic waves travel
- understand what is meant by frequency and wavelength, and how these relate to the speed of a wave
- recall that all electromagnetic waves travel at the same speed in a vacuum.

Light is a type of **electromagnetic radiation**. These radiations are disturbances in electric fields. They move energy from one place to another by travelling as waves. Electromagnetic waves don't need to vibrate particles to travel. This means that they can move easily through a **vacuum**, such as space. So easily, that they are the fastest thing there is, travelling at 300 000 kilometres every second. An aeroplane going at this speed could go around the world more than seven times in one second.

Radio waves, microwaves, infra red, ultraviolet, X-rays and gamma rays, as well as visible light, are all types of electromagnetic radiation.

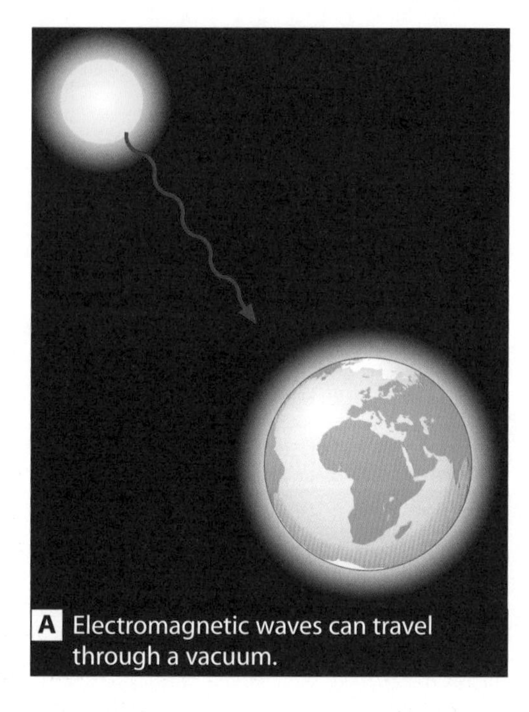

A Electromagnetic waves can travel through a vacuum.

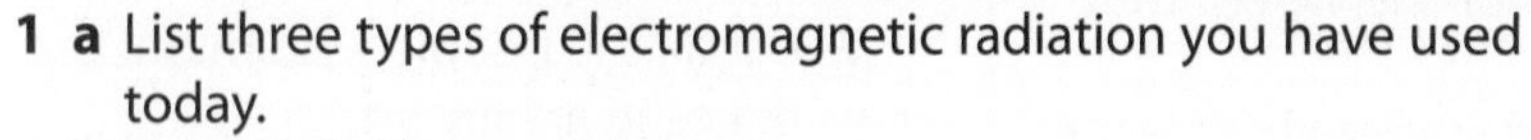

B There are many types of electromagnetic wave.

1 a List three types of electromagnetic radiation you have used today.

b What did you use them for?

2 How do you know electromagnetic radiation can travel through a vacuum, such as space?

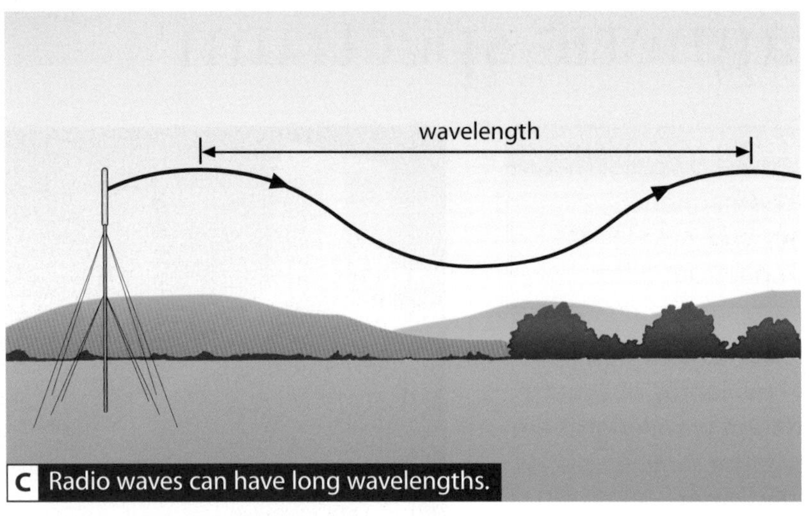

C Radio waves can have long wavelengths.

All electromagnetic waves can be measured in various ways. The **wavelength** (λ) is the distance from one point on the wave to the same point on the next wave. Electromagnetic waves can range from several kilometres (radio waves) down to thousand millionths of a millimetre (gamma rays).

The **frequency** of a wave is the number of complete vibrations per second. Again, electromagnetic waves have a huge range of possible frequencies. For example, microwaves can have a frequency of a thousand million waves per second. The standard unit for frequency is **hertz** (**Hz**); 1 hertz is one complete wave per second.

Electromagnetic waves follow the wave equation.

wave speed (metres/second, m/s) = frequency (hertz, Hz) x wavelength (metres, m)

$v = f \times \lambda$

Example

BBC Radio One broadcasts at 98.1 MHz which means 98 100 000 Hz. What is the wavelength of this radio wave?

frequency = 98 100 000 Hz, speed = 300 000 000 m/s

$v = f\lambda$

$\lambda = v/f$

$\lambda = 300\,000\,000/98\,100\,000$

$= 3.058$ m

$\lambda = 3.06$ m

Possible effects of mobile phones on people

We don't clearly understand the effects of some types of electromagnetic waves on humans. Some people are worried that mobile phones may cause cancer. Government scientists have not yet found evidence for this, but research is still being done.

3 What does it mean to say that an X-ray has a frequency of a million million million Hertz? P

4 Natalie noticed some information on the back of her TV remote controller: 'Infra red waves: $\lambda = 0.001$ m'. What does this mean? V

5 What is the speed of X-rays in a vacuum? P

6 Virgin Radio broadcasts at a frequency of 1215 kHz which means 1 215 000 Hz. What is the wavelength of this radio wave? Show your working. (*Hint*: you will need to rearrange the equation $v = f\lambda$.) P

7 The wavelength of orange light is 0.000 0006 m, and its frequency is 500 000 000 000 000 Hz. Calculate the speed of orange light. Show your working and explain any symbols you use. Make sure you use units in your calculation. D P P

P1b.2

The electromagnetic spectrum

By the end of this topic you should be able to:

- understand what is meant by the electromagnetic spectrum
- explain how the different parts of the spectrum differ.

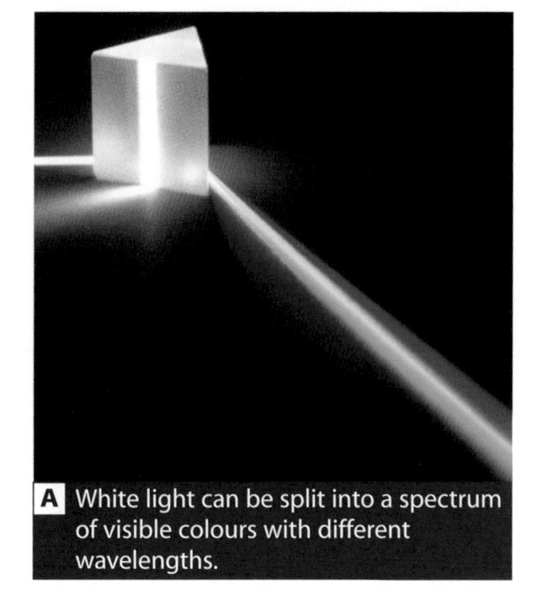

A White light can be split into a spectrum of visible colours with different wavelengths.

Light has a wavelength which can be measured in metres, although it is very small. It ranges from 0.4 millionths of a metre to 0.7 millionths of a metre. Light with different wavelengths are different colours. All colours of light travel at the same speed. As the wavelength increases, the frequency decreases.

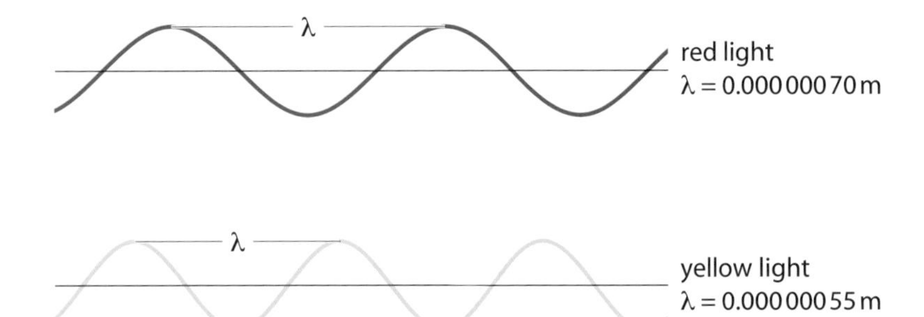

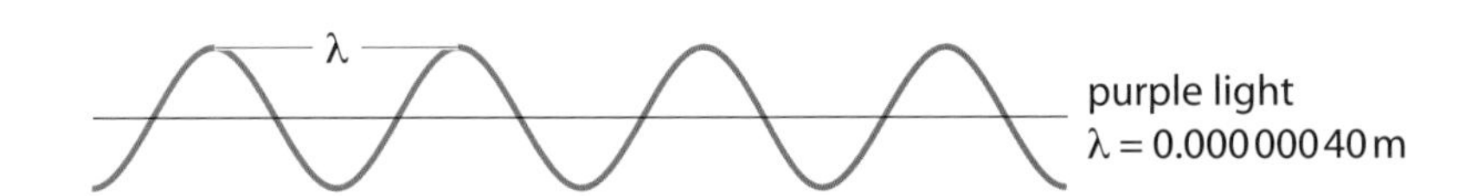

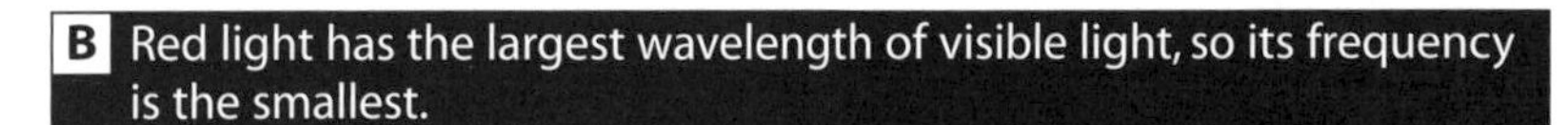

B Red light has the largest wavelength of visible light, so its frequency is the smallest.

1 What is the difference between red and yellow light, in terms of their:
a wavelength
b frequency
c speed?

Many parts of the electromagnetic spectrum are used in an airport: from radio waves for walkie-talkies to the X-rays used in luggage scanners. Some types of electromagnetic wave can be dangerous, so in those cases the equipment includes protection for the users. For example, the airport X-ray machines that check luggage have lead casings to stop the X-rays escaping out into the general environment.

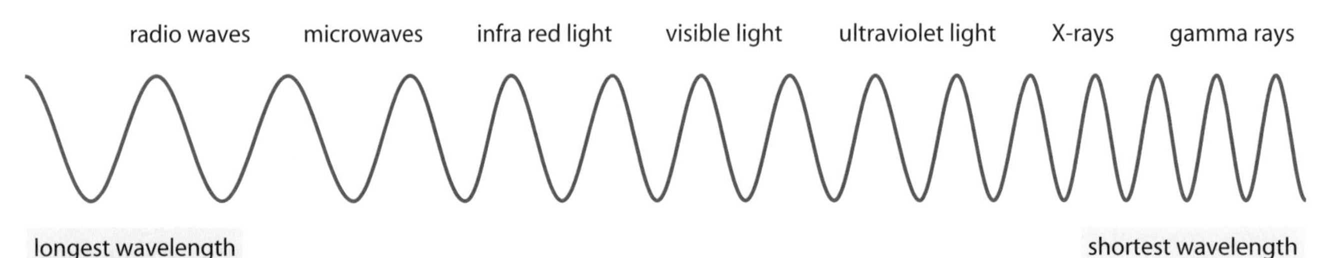

C The electromagnetic spectrum.

The colour of a light wave depends on its wavelength. The colours of the rainbow that we can see are known as **visible light**. If the wavelength of a light wave is longer than that of red light, human eyes cannot see it. It is still there though and is called **infra red (IR)**. **Microwaves** and **radio waves** have longer wavelengths again. Radio waves can be anything from 50 centimetres up to hundreds of kilometres in wavelength.

If the wavelength of a light wave is shorter than that of purple light, it is called **ultraviolet (UV)**. It has a higher frequency than visible light. **X-rays** have even shorter wavelengths than UV and **gamma rays** shorter still.

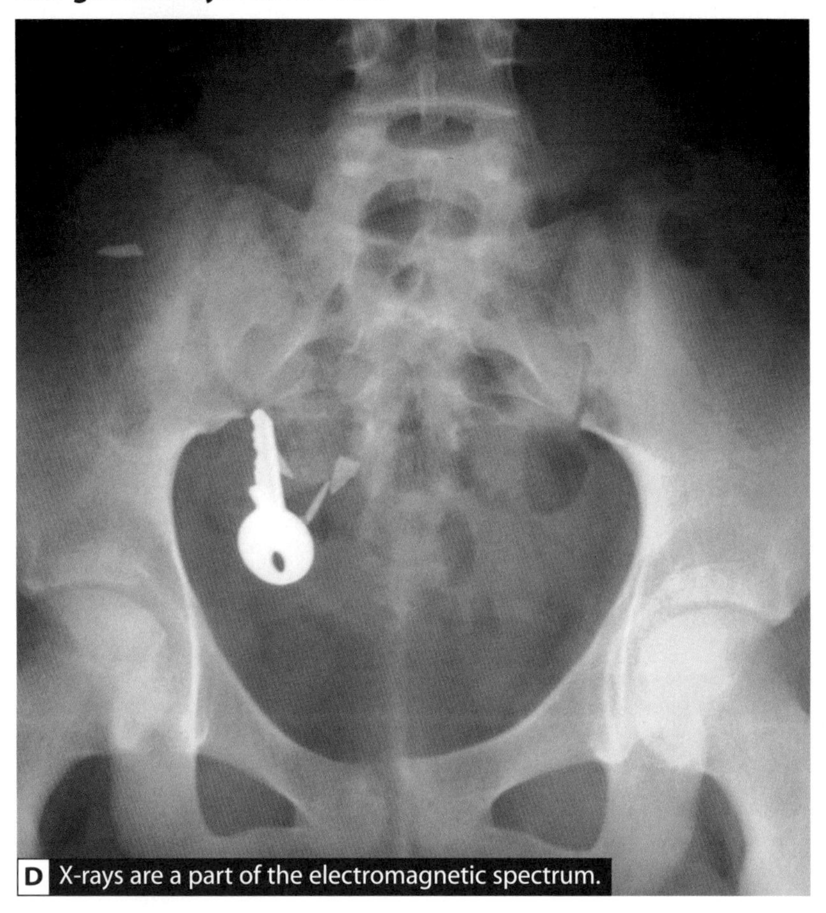

D X-rays are a part of the electromagnetic spectrum.

The full range of electromagnetic waves is called the **electromagnetic spectrum**. It's like a bigger version of a rainbow and includes all the wavelengths we cannot see. The spectrum is continuous, but the wavelengths can be grouped into seven types of which visible light is only one. Some of the different parts of the electromagnetic spectrum are useful, some are harmful.

5 Which part of the electromagnetic spectrum has a higher frequency than X-rays?

6 If an electromagnetic wave has a wavelength between visible light and X-rays, what type of wave is it?

2 Name three different types of electromagnetic wave.

3 Give two ways in which radio waves and microwaves are different.

4 What is the largest wavelength that human eyes can sense?

7 Draw a diagram to show the electromagnetic spectrum, including labels for all seven parts of the spectrum in the correct order. Add arrows on the diagram to show increasing wavelength and increasing frequency.

P1b.3

Radio waves

By the end of this topic you should be able to:

- describe what radio waves can be used for
- explain what happens when radio waves meet different substances, including living cells and metals
- describe the hazards of radio waves.

B Radio messages from the control tower help planes to land.

The longest wavelengths in the electromagnetic spectrum are radio waves. Any electromagnetic wave with a longer wavelength than 50 cm is a radio wave.

BA 871 cleared for take off.
1 Sound changed into electrical signals by microphone.
2 Electrical signals changed into radio signals by transmitter.
3 Radio signals changed back to electrical signals by receiver.
4 Loudspeaker changes electrical signals back into sounds.
BA 871 cleared for take off.

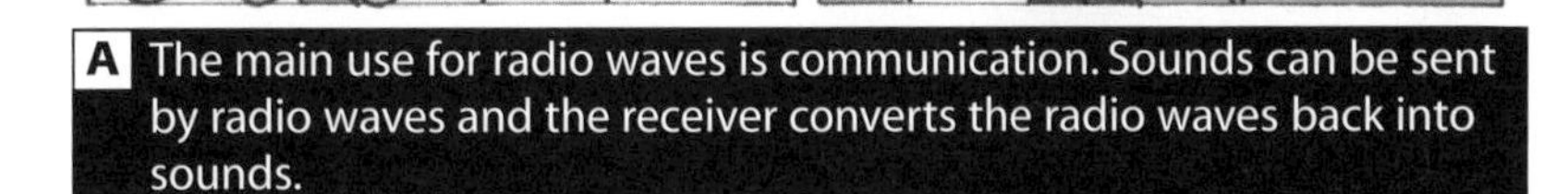

A The main use for radio waves is communication. Sounds can be sent by radio waves and the receiver converts the radio waves back into sounds.

1 Where do we find radio waves in the electromagnetic spectrum?

2 a Give two uses of radio waves.
b For each use in part **a**, explain why radio waves are used.

Radio is now so widely used that our planet is constantly awash with radio waves. When radio waves hit living cells, they mostly pass through them. This means that nothing happens, so we think it is a safe technology to use. However, high-energy radio waves can be absorbed by tissues, which causes them to heat up. This may cause damage such as cataracts in the eyes. For example, being very close to a **radar** antenna can be dangerous, so workers are not allowed near them when they are switched on. A walkie-talkie has such a low power that it is completely safe.

C Strong radio waves can be dangerous, but are not common.

Radio waves are **absorbed** by thicker or denser objects like thick walls. Metals can absorb or **reflect** them. For a receiver to pick up radio signals, the aerial must be made of metal so it absorbs the waves. When it does so, the radio waves produce an alternating current in the aerial. The frequency of this current is the same as the radio wave itself. This electric signal can be changed into the sound that comes from a loudspeaker. The electric current in the metal may also make it warmer.

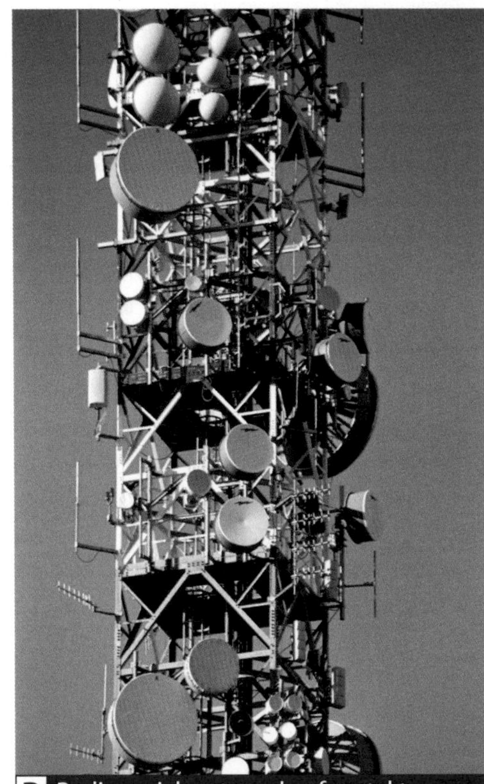

D Radio aerials are made of metal to absorb the waves.

3 a Why would a walkie-talkie have a limited range inside an airport?

b Why would its range increase outside on the airfield?

c How would the number of people in an airport affect walkie-talkie range?

4 How is alternating electric current involved in receiving radio waves?

5 Why should workers on the airfield at Heathrow Airport keep away from the radar antennae?

Radio waves are made naturally by some stars. Astronomers study these using banks of big dish receivers which can pick up these radio waves. To pick up very large wavelengths we can link together the results from several radio dishes which can be many kilometres apart.

The largest single-dish radio receiver in the world is at Arecibo in Puerto Rico. One of its uses is to look for radio messages from space that may have been sent by aliens. This forms part of the Search for Extra-Terrestrial Intelligence (SETI) project. So far, no alien messages have been received; although one unusual one from 1977 remains unexplained.

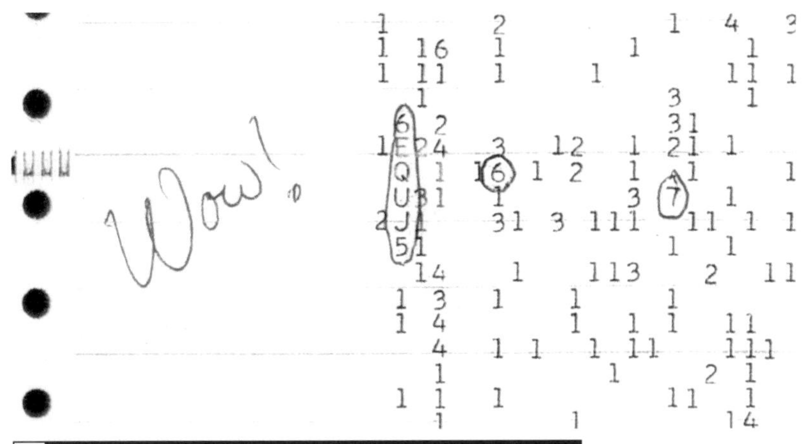

E Is the 'Wow' signal a radio message from aliens?

6 Where are radio waves made naturally?

7 a What does SETI stand for?

b What is the aim of SETI?

8 Draw a diagram to show how two astronauts on the Moon could communicate using radios built into their spacesuits. Include labels to show the role of metals in the communication.

Microwaves

By the end of this topic you should be able to:

- describe the uses of microwaves including their use in communication
- explain the dangers of microwaves.

Short wavelength electromagnetic waves of between 1 mm and 50 cm are called **microwaves**.

Mobile phones on planes

Most airlines do not allow mobile phones to be used on aeroplanes. This is because the microwaves they use might interfere with the electronic equipment. There is not enough evidence to prove or disprove this as yet.

1 Why might it be dangerous to use your mobile phone on a plane?

The ionosphere is a high and electrically-charged layer of our atmosphere. Many radio frequencies are reflected by the ionosphere. This means they cannot be used to communicate with satellites. Microwaves are used instead as these can pass easily through the whole atmosphere.

2 Why can't radio waves be used to signal to a satellite?

A Satellite to Earth signals are sent using microwaves.

Mobile phones also work using microwaves and a cellular network system. The phone communicates with a local mast or base station. This sends another microwave signal to a central base which sends the signal on to the base station for the receiving phone.

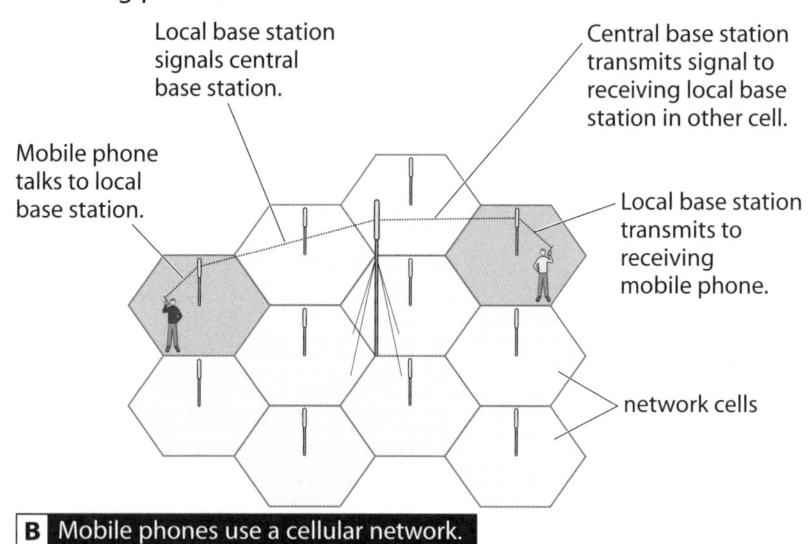

B Mobile phones use a cellular network.

For long distances, mobile phone microwave signals are sent via several masts as the curved shape of the Earth would block the signal. For international calls, a satellite link is used.

Mobile phones and health

There have been public worries about the effects of mobile phones and their masts on our health. The Health Protection Agency (HPA) is a government group, made up of independent scientists, which is looking at the evidence about the effects of radiation on health. Their opinion is that the microwaves used in mobile phones do not have any health risks.

C Microwave phone signals are relayed by booster masts.

3 **a** What does HPA stand for?
   b What does the HPA do?

4 Which part of the electromagnetic spectrum is used by mobile phones?

5 Describe what a cellular network system, such as that used in mobile phone networks, is.

6 Why are satellites used for international calls?

D Microwaves can be used to cook food.

Microwaves are a form of energy. Those with a frequency of 2.45 thousand million hertz can be absorbed by water molecules. This gives the molecules more energy and the water heats up. So, any food containing water can be cooked by microwaves. This could also be dangerous as our bodies are mostly water. The waves are absorbed by metal so microwave ovens have metal sheets or mesh around the cooking area to protect people. Mobile phones do not use the same frequency as microwave ovens. This is one reason why the HPA say that there is no known danger in using them.

7 Explain how a cup of cold tea is reheated by a microwave oven.

8 A chef in Paris once removed the doors from the restaurant's microwave ovens for speedy use. Why was this dangerous?

9 Explain why the microwaves used in microwave ovens are dangerous, but those used in mobile phones are not.

10 Using a diagram, explain how a mobile phone allows you to:
   a call a friend in Australia
   b call a friend in a nearby town.

Infra red radiation

By the end of this topic you should be able to:

- explain the uses of infra red radiation, including why it is used in optical fibres
- describe the hazards of infra red radiation.

Infra red radiation (**IR**) is electromagnetic radiation with a wavelength between that of microwaves and visible light. It has wavelengths between 0.7 millionths of a metre and 1 millimetre.

IR travels well inside glass and clear plastic, but is absorbed by dark surfaces and reflected by shiny surfaces. These properties are very similar to those of light, but we cannot see infra red.

1 What is the shortest wavelength for IR waves?

2 How fast do infra red waves travel through space?

3 In what ways is IR:
a similar to visible light
b different from visible light?

When you check in at an airport, your information is sent from the check-in computer to another at the departure gate. The information is coded in infra red radiation and travels along an optical fibre.

Infra red signals sent down **optical fibre** cables have transformed telephone and internet communications. A wave property called **total internal reflection** allows light and IR to travel in a very thin 'light pipe' made of glass or plastic.

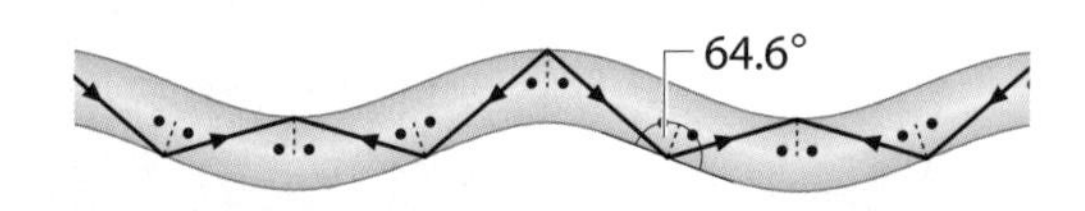

A IR can be guided by a 'light pipe' to pass on signals.

Total internal reflection only happens if a wave tries to leave a dense medium for a less dense one; for example, light leaving a piece of glass to go into the air. At large angles of incidence, the waves do not escape and are reflected back inside.

Angle of incidence is less than the critical angle.

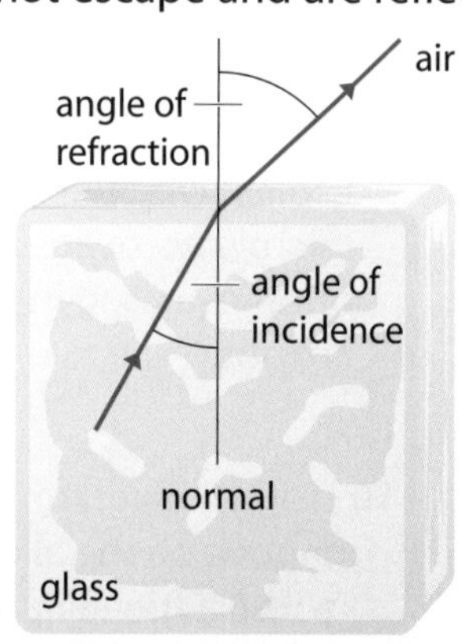

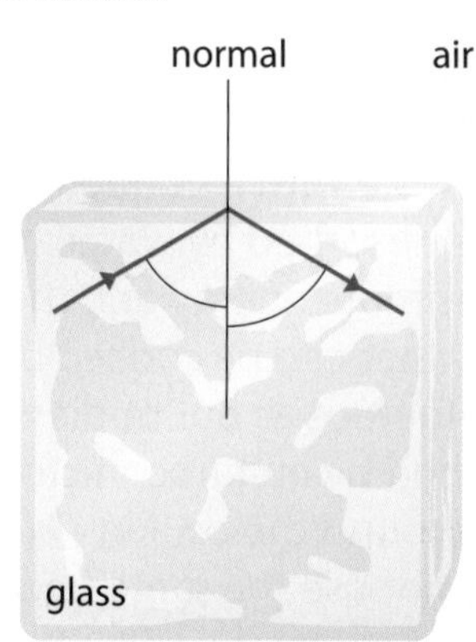

Angle of incidence is more than the critical angle.

B At large angles of incidence, light and infra red will be totally internally reflected in glass or plastic.

A long narrow tube of glass can keep reflecting light, or IR, right along its length. This is an optical fibre and we use them to carry signals for many kilometres. Optical fibre communications are very fast and are free from interference.

Infra red is also known as heat radiation. It is absorbed by soft tissue, causing it to heat up. Too much infrared can cause burns. It is used for cooking in grills and toasters.

All warm objects give off some heat radiation. This travels through fog and in the dark. This means cameras sensitive to it can 'see' in the dark.

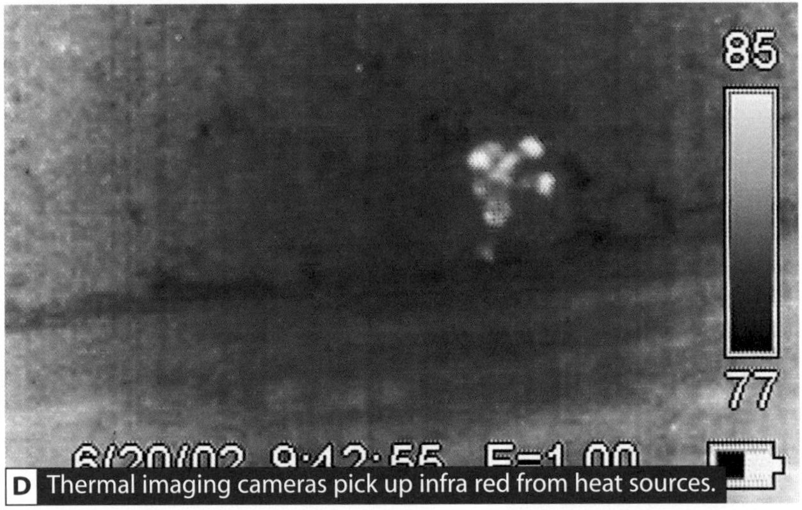

D Thermal imaging cameras pick up infra red from heat sources.

CCTV cameras can be used at night if they are built to 'see' IR as well as visible light. Infra red heat sensing is also used in burglar alarms.

6 What properties of the Sun show that it gives off infra red radiation?

7 Why do police helicopters often have infra red cameras?

8 Write one benefit and one problem if your eyes could see infra red radiation.

4 Draw a diagram to show how an infra red signal could be sent between computers in different parts of an airport.

5 Look at diagram A and suggest why this is known as 'total internal reflection'.

C Cooking with IR.

9 **a** List five uses for IR.
b For each use given in part **a**, explain briefly how it works.

P1b.6

Visible light

By the end of this topic you should be able to:

- describe the uses of visible light.

Visible light makes up a very narrow band in the middle of the electromagnetic spectrum. It includes wavelengths from 0.4 to 0.7 millionths of a metre. It is visible to us as the colours in the rainbow. Everyday life shows us that it will pass through clear materials, like glass. We say glass **transmits** visible light. It is absorbed by dark, dull materials. Light, shiny surfaces reflect visible light.

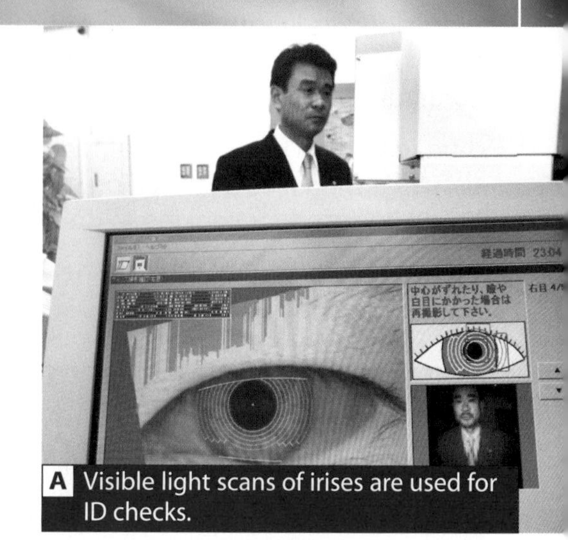

A Visible light scans of irises are used for ID checks.

1 Copy these sentences, adding in the missing word in each case. Choose from: absorbed, transmitted, reflected.

a Visible light is ________________ by glass.
b Visible light is ________________ by brick.
c Visible light is ________________ by polished silver.

Visible light has long been used in communications. You are reading this book by seeing the visible light which enters your eyes.

B Visible light guides aeroplanes in to land.

As with infra red, visible-light wavelengths can be carried in optical fibres by total internal reflection. This is used in computer networks and **telecommunications**.

C Optical fibres can carry huge amounts of information quickly and accurately.

2 List five ways people might send messages using visible light.

3 Give one benefit and one problem with optical fibres.

The most common and obvious use for visible light is to look at things. **CCTV** cameras and mobile phones with pictures are so common now that much of what we do is recorded for us to look at again later.

D Airport security is improved by recording images of all passengers.

Privacy

Many people are worried that it is not possible to have any privacy as so much that we do is recorded by cameras.

4 How could CCTV be useful at home?

5 Cameras can now be made so small that they can hardly be seen. How might this become a problem for people's privacy?

Visible light is used by doctors in an endoscope. This device has an optical fibre cable which is passed into the patient so the doctor can see internal problems.

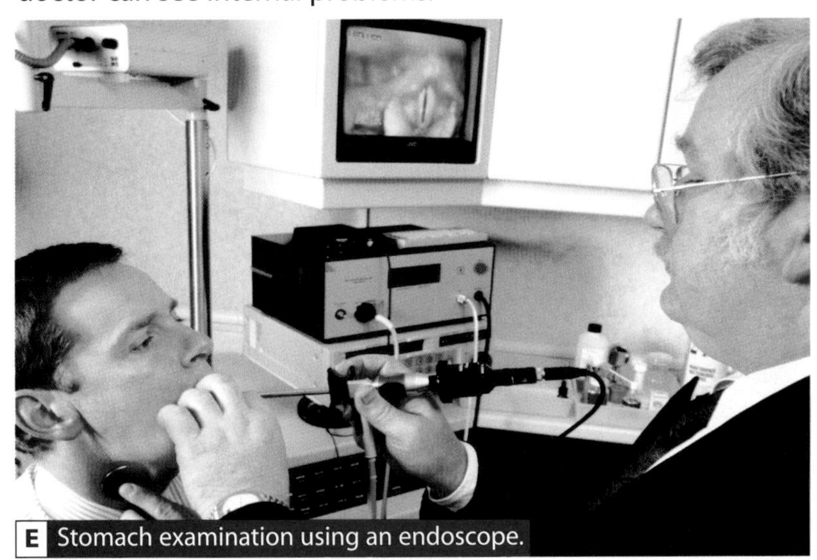

E Stomach examination using an endoscope.

6 An endoscope contains more than one optical fibre. How could this overcome the problem that it is dark inside a patient?

The low-energy half of the electromagnetic spectrum includes radio waves, microwaves, IR and visible light.

F Electromagnetic waves are used on aeroplanes.

7 Look at photograph F.

a List the ways light is being used by the pilot.

b List all of the ways low-energy electromagnetic waves (radio, microwaves, infra red and visible light) are being used on the aeroplane.

Ultraviolet radiation

By the end of this topic you should be able to:

- describe the uses of ultraviolet radiation
- explain the hazards of ultraviolet radiation
- describe how ultraviolet radiation behaves when it hits materials.

As the frequency of electromagnetic waves increases above that of visible purple light, we get **ultraviolet (UV)** light. Human eyes cannot see ultraviolet, but birds can. As electromagnetic frequency increases, the energy carried by the waves increases. This means UV rays carry more energy than visible light. Like visible light, UV is absorbed by dark surfaces, transmitted by clear materials, and reflected by shiny or metal surfaces.

One of the main uses for UV light is in simple security measures. Some inks are invisible to humans, but when you shine ultraviolet light on them, they absorb the UV energy and emit it as visible light, and so we can see the ink. This is used on many banknotes to prove they are real. The same system is also used to mark bikes and other property in case of theft.

A Property can be security marked with a cheap UV pen.

1 Describe where ultraviolet radiation is in the electromagnetic spectrum.

2 a Give two ways in which UV is similar to visible light.
  b Give two ways in which UV is different from visible light.

B Some flowers reflect UV in patterns to attract insects that can see it.

3 How can we see the hidden markings on a banknote?

4 Explain how a theme park could check if people have paid, using UV security ink on customers' hands.

Ultraviolet radiation carries a lot of energy, and this can be dangerous to us. Have you ever been sunburnt? Sunburn is caused by over-exposure to UV energy.

Your skin can absorb UV rays. The energy of the UV radiation then heats up your cells. Too much exposure to ultraviolet radiation can provide so much heat that the skin burns. Over a long time, this can also damage the skin cells and cause skin cancer.

Skin naturally has some defence against ultraviolet radiation. Fair skin usually produces a substance called melanin when it is exposed to the Sun. This increases the natural protection against the harmful UV radiation. Suncream, clothing and a hat, can also increase protection against the harmful rays by absorbing the UV before it reaches the skin.

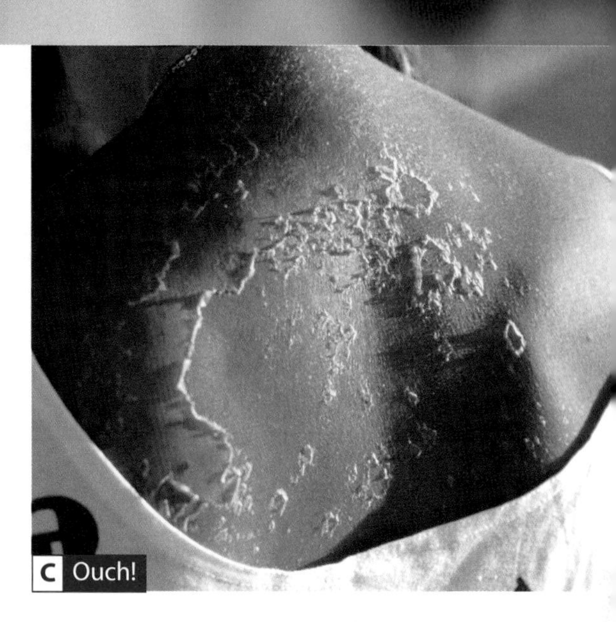

C Ouch!

D Suncream blocks UV light. The higher the sun protection factor (SPF) number, the more protection it gives.

5 Give two ways to protect yourself against skin damage by UV radiation from the Sun.

6 Why do people who are exposed to the Sun a lot develop darker skin?

7 **a** What are the dangers to skin of too much exposure to the Sun?

b Explain your answer in terms of skin cells and energy.

8 In photograph D, how is the child protected against the Sun?

9 Draw a spider diagram to link the following words:
ultraviolet radiation, UV, sunburn, suntan, suncream, SPF number, security, skin cancer, sunglasses, electromagnetic spectrum, energy, banknotes.
Add labels to explain the links between the words.

V

P

D

P

P

X-rays

By the end of this topic you should be able to:

- describe the uses of X-rays
- give examples of the hazards of X-rays
- explain how X-rays behave when they come into contact with other materials.

X-rays come between UV and gamma rays in the electromagnetic spectrum. Wilhelm Röntgen discovered X-rays in 1895. He had no idea what he had found, so he called it 'X radiation'.

1 How do the following properties of X-rays compare with ultraviolet radiation:
  a frequency
  b wavelength
  c speed?

A Airport security scanners show up metal objects that absorb X-rays.

Airport scanners work by sending a beam of X-rays into an item of luggage. As they carry a lot of energy, these electromagnetic waves pass through many solid materials without being absorbed, and are picked up by a detector underneath the luggage. However, X-rays are absorbed by metals. So any metal objects inside the luggage will show up on the detector as a gap in the shape of the object.

2 **a** Name one material X-rays can pass through.
  b Name one material that will absorb X-rays.
  c Explain how this makes it useful for airport security.

The same idea can help companies which use metal parts. A metal object placed in a scanner absorbs X-rays and the detector can make very accurate measurements of the size and shape of it. X-rays can also be used to show up faults, even a tiny crack, in machinery or pipes that are already in use.

Many people have had an X-ray photograph taken if they have broken a bone. Bone is dense enough to absorb X-rays, but flesh is not. When X-rays are sent through the patient there is a photographic film on the other side. The film becomes cloudy in areas where X-rays get through, i.e. under areas with no bones. This creates a shadow picture of the bones. Doctors use these to see inside patients without cutting them open.

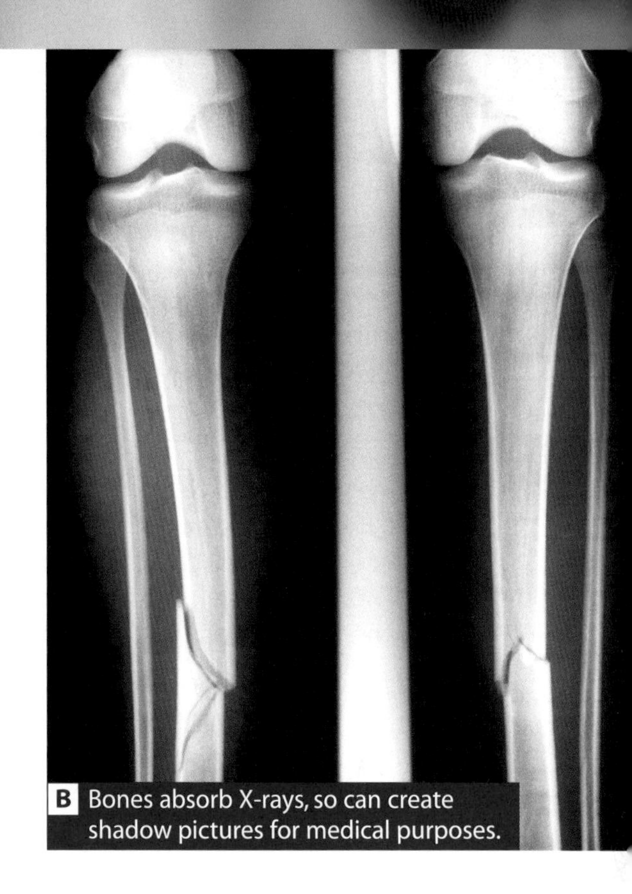

B Bones absorb X-rays, so can create shadow pictures for medical purposes.

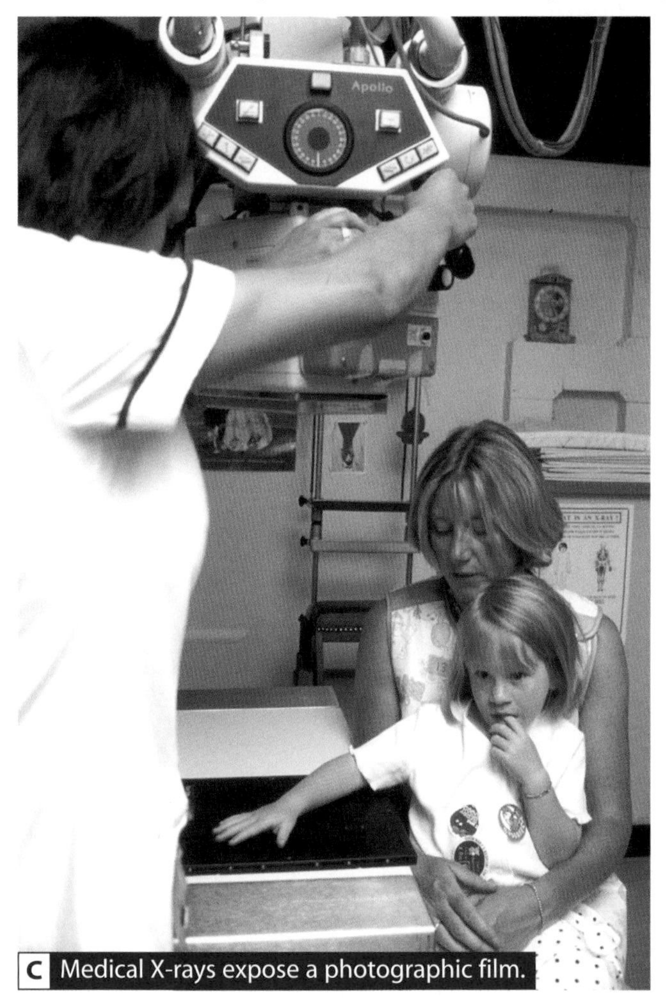

C Medical X-rays expose a photographic film.

Care must be taken as X-rays can be dangerous. The radiation carries a lot of energy, and this can damage our body. Although it mostly passes through the body, some parts of the body can absorb the X-rays. The cells in these areas may become damaged if they are exposed too much. Large doses can eventually cause cancerous changes and may kill cells. Children and babies are most at risk, so these scans are rarely used on pregnant mothers. In small doses, for example two chest X-rays per year, the body can recover fairly quickly.

3 Give one danger from X-rays.

4 Why is an X-ray still a good idea for many patients even though they are dangerous.

5 Why could an X-ray not be used to diagnose a damaged lung?

6 X-rays are especially damaging to cells that are dividing. Why might children and babies suffer more from having an X-ray than adults?

7 A doctor has asked you how airport security scanners could discover his surgical knives. Write an explanation for the doctor. In your answer, compare the luggage scan with a medical scan.

Gamma rays

By the end of this topic you should be able to:

- describe the uses of gamma rays
- describe the hazards of gamma rays
- explain how gamma rays behave when in contact with materials.

The highest frequency electromagnetic waves are called **gamma rays**. As they have the highest frequency, they carry the most energy. They can pass through anything. Very dense or thick materials, like lead, can absorb some of their energy. Taking this energy from the gamma rays makes them safer.

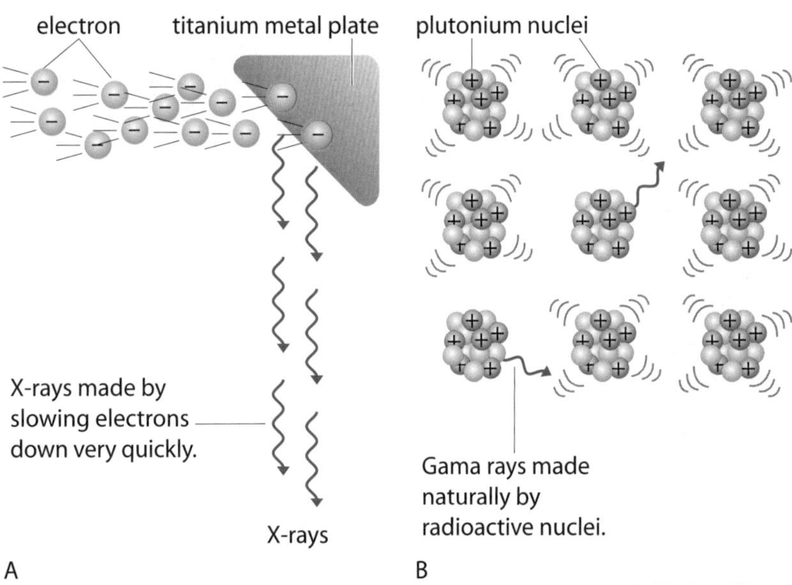

A Some X-rays and gamma rays are the same, but they are made differently.

There is an overlap between the lower frequency end of gamma rays and the upper frequency end of X-rays. In the overlap area, both types of electromagnetic wave are exactly the same. The difference is in how they are made. A **radioactive nucleus** sends out a gamma ray making the nucleus more stable. X-rays are made by slowing **electrons** down very quickly. In an X-ray machine, this is done by firing a beam of fast-moving electrons into a metal plate. The plate stops the electrons so rapidly they give off X-rays.

Gamma rays are dangerous. If their high energy is absorbed by a cell, it may cause cancerous changes in the cell or kill it. However, gamma rays are sometimes used to treat cancer. This is because they can be used to kill the cancer cells directly. This **radiotherapy** has to be undertaken with great care so as not to cause further cancers, in either the patient or hospital staff.

1 Where does gamma radiation fall in the electromagnetic spectrum in terms of frequency?

2 How does the frequency affect the amount of energy carried by an electromagnetic wave?

3 Why are gamma rays more dangerous than any other type of electromagnetic wave?

4 Some gamma rays have the same frequency as some X-rays. Give a difference between these two identical electromagnetic waves.

Gamma radiation can also be used for medical scans of some soft tissues. A radioactive chemical is injected into the blood. The gamma rays given off can be detected by a gamma camera, so the flow of blood can be watched to check for problems.

The gamma camera can be used to detect medical problems such as spine damage. Again, great care must be taken and only a tiny dose of radioactive chemical is injected to make sure the patient does not get sick from the test itself. As gamma rays pass easily through the human body, only a small amount is needed.

5 Describe how gamma rays can be used for medical purposes.

6 Why must doctors be careful when using gamma-ray equipment?

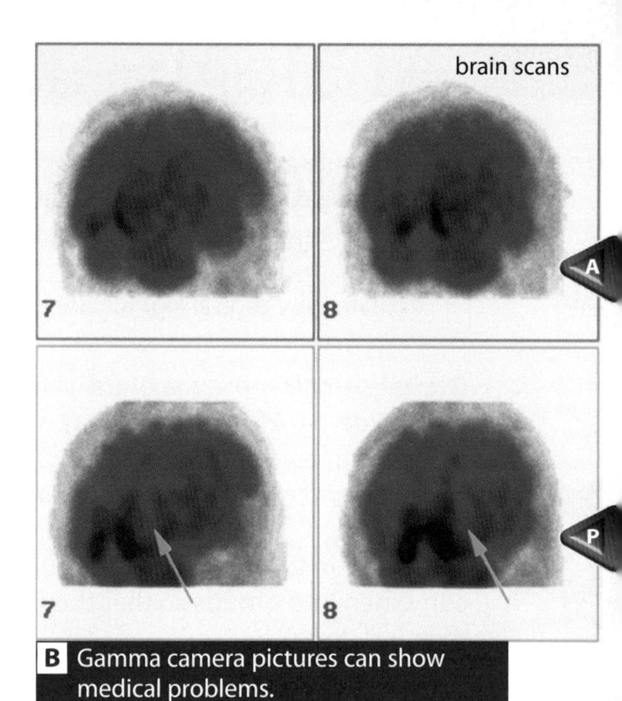

B Gamma camera pictures can show medical problems.

Gamma rays can also be used to kill harmful bacteria in food and on surgical instruments. As they have such high energy, the waves kill the bacteria cells. This is called sterilisation. In gamma-ray equipment, lead shielding is used to protect the user by absorbing most of the energy of escaping gamma rays.

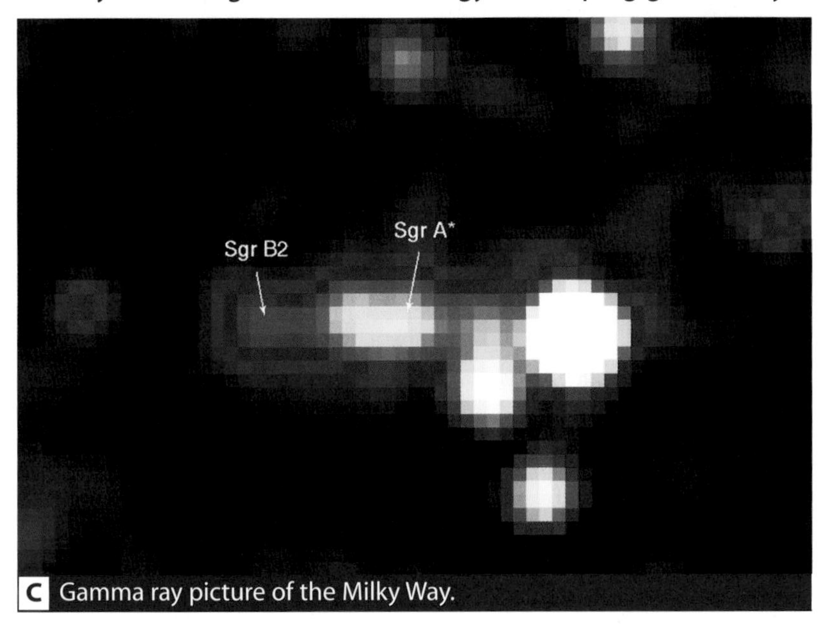

C Gamma ray picture of the Milky Way.

Gamma-ray astronomy

Astronomers can view the sky using different equipment to pick up different parts of the electromagnetic spectrum. Many types of star give off gamma radiation and **gamma-ray astronomy** is an important study.

7 Why would gamma-ray astronomers have no problems with the weather?

8 **a** Write down the seven types of electromagnetic wave in order, from the longest to the shortest wavelength.
b Give one use for each type.
c Show how the frequency changes across the spectrum.
d Show how the wavelength changes across the spectrum.
e Describe how each type behaves in contact with the other materials.
f Show how the dangers change across the spectrum.

Analogue and digital signals

By the end of this topic you should be able to:

- describe the difference between analogue and digital signals
- explain why digital signals are better for communications.

Digital signals are some of the simplest we can send. They can only be **on** or **off**. In a similar way a torch bulb can be flashed on and off; for example, three flashes on might mean 'help!'. These on and off signals are sometimes called **high** and **low**, or **1** and **0**. Each signal, high or low, 1 or 0, is called a bit. Bit is short for 'binary digit'. To send messages, bits need to be combined into groups so that they can give more possibilities than just 1 or 0.

1 What is the word 'bit' short for?

2 The two possibilities for a bit can be described as 'high' or 'low'. Give two other ways of describing them.

3 How are the bits used to be able to send a message?

A	00001	**L**	01100	**W**	10111
B	00010	**M**	01101	**X**	11000
C	00011	**N**	01110	**Y**	11001
D	00100	**O**	01111	**Z**	11010
E	00101	**P**	10000	*spare*	11011
F	00110	**Q**	10001	*spare*	11100
G	00111	**R**	10010	*spare*	11101
H	01000	**S**	10011	*spare*	11110
I	01001	**T**	10100	*spare*	11111
J	01010	**U**	10101	*spare*	00000
K	01011	**V**	10110		

A A digital code for the alphabet.

4 Use the code in table A to write your name in digital code.

5 Suggest a way that you could send this as a digital message.

Analogue signals can have many different possible values. In an airport loudspeaker announcement, the sound of the voice can vary between many different loudness levels. Having an unlimited number of levels is what 'analogue' means, and a signal which changes in exactly the same way as the information it is carrying is called an 'analogue signal'.

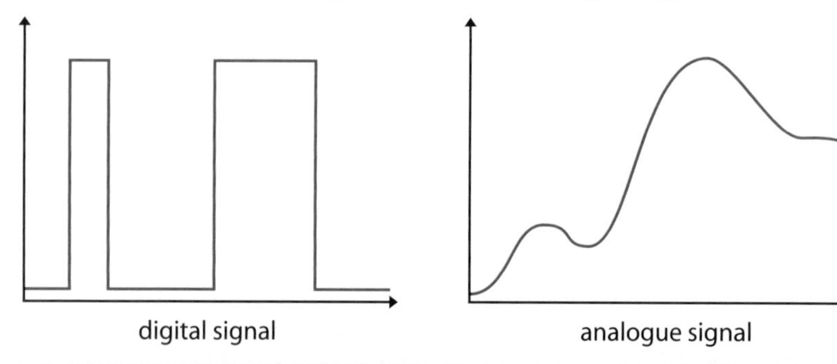

B A digital signal is only high or low, an analogue signal has many possible levels.

All signals suffer from **noise**, which is **interference** or random changes picked up as they travel. The difference between the information sent and the information received can cause problems. Digital signals generally have less of a problem with this. When a digital signal arrives, it is easy for the receiver to work out when the signal should be high and when it should be low. With an analogue signal it is not so obvious that the received signal may have changed slightly from what was sent.

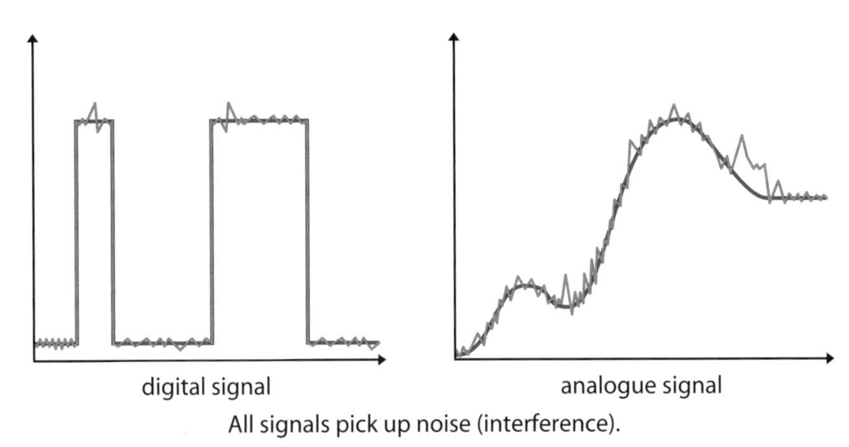

All signals pick up noise (interference).

C Digital signals can easily have interference identified.

If a radio station is broadcasting music, interference may make the music sound poor. With digital signals the receiver can ignore the interference by still seeing each pulse as either high or low. So, the string of 1s and 0s will be received correctly, and the signal will be identical to that sent. As computers can process digital signals, they suffer fewer problems from interference, so emails don't arrive with mistakes in them.

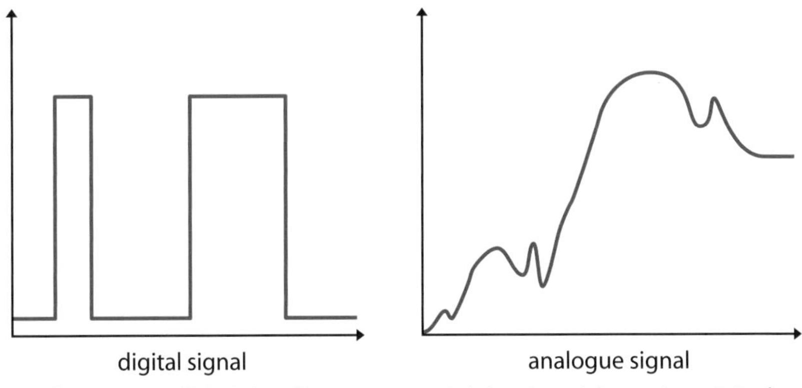

Interference on a digital signal is easy to spot and so can be easily removed.

It is hard to pick out the original analogue signal from the noisy version.

D Analogue signals suffer more from interference than digital ones.

6 Which type of signals pick up the most noise?

7 Why does noise cause fewer problems to digital signals?

8 Why do computers suffer less from interference than radios?

9 Write a short newspaper article to explain to readers why most communications, including mobile phones, radio and TV, are changing from using analogue to digital signals. Include a diagram to show why interference is less of a problem with digital signals.

Where does radioactivity come from?

By the end of this topic you should be able to:

- describe what an atom is made from
- explain what an isotope is
- explain what it means for an atom to be radioactive.

We know that all objects are made from **atoms**. Atoms are thought of as the smallest building blocks of materials. In 1911, Lord Rutherford came up with an idea for how an atom itself was made up.

An atom has a small **nucleus** (plural = nuclei) in the middle. This is made from two types of tiny particles: **protons** and **neutrons**. Protons are positively charged, and neutrons have no charge. Most of the atom is just empty space. Electrons orbit the nucleus towards the edge of the atom. These have the same size charge as the protons, but their charge is negative. Electrons are even smaller than protons and neutrons. Atoms have zero charge overall, so they always have the same number of protons and electrons.

1 Look at diagram B. How many of the following would you find in a single carbon atom?
 a electrons
 b neutrons
 c protons

The number of protons an atom has is called its **atomic number**. If you add the number of protons and the number of neutrons, you get the **mass number** for that atom. **Isotopes** are atoms with the same atomic number but different numbers of neutrons. Isotopes have different mass numbers.

Some atoms have unstable **nuclei**. This is because the mix of protons and neutrons is not quite right and the nuclei need to change. We say that they **decay** to become more **stable**. The change involves giving off radiation, so these elements give out radiation until they become stable. Some of these elements, like radon gas, are present around us all the time. We are exposed to a low level of **background radiation** all the time. It is always there and will add to readings in radioactivity experiments. This means that you must take away the background count for any radioactivity readings you take.

2 If a positive and a negative charge cancel each other out, what is the total charge on a carbon atom?

A An aeroplane is made from billions of tiny atoms.

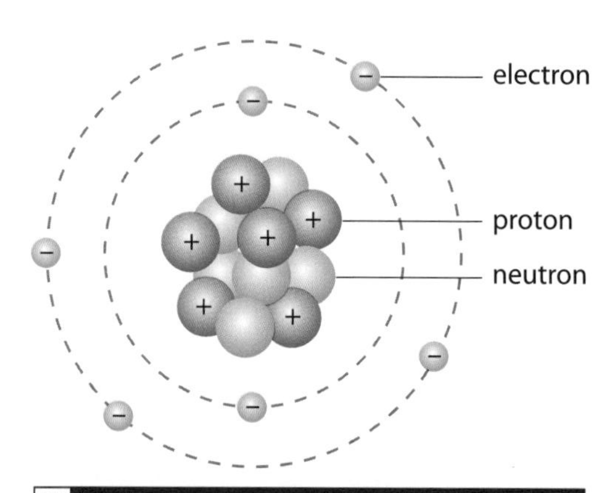

B Rutherford's idea of a carbon atom in close-up. The atom is about 0.1 thousand millionths of a metre across.

Particle	Mass	Charge
Electron	almost zero	−1
Neutron	1	0
Proton	1	+1

C Charge and mass properties of the parts of atoms.

3 An oxygen atom has eight protons.
 a How many electrons does it have?
 b What is the overall charge of an oxygen atom?

4 Write a rule for the number of protons and electrons in any atom.

5 Why do some elements give off radiation?

6 Kimberley did an experiment to measure how radioactive a brazil nut is. Her reading was 4 counts per second. Amira pointed out that she had not corrected for the background radiation which was 0.5 counts per second. What was the correct reading for the brazil nut?

D Nuclei of radioactive atoms will give off radiation.

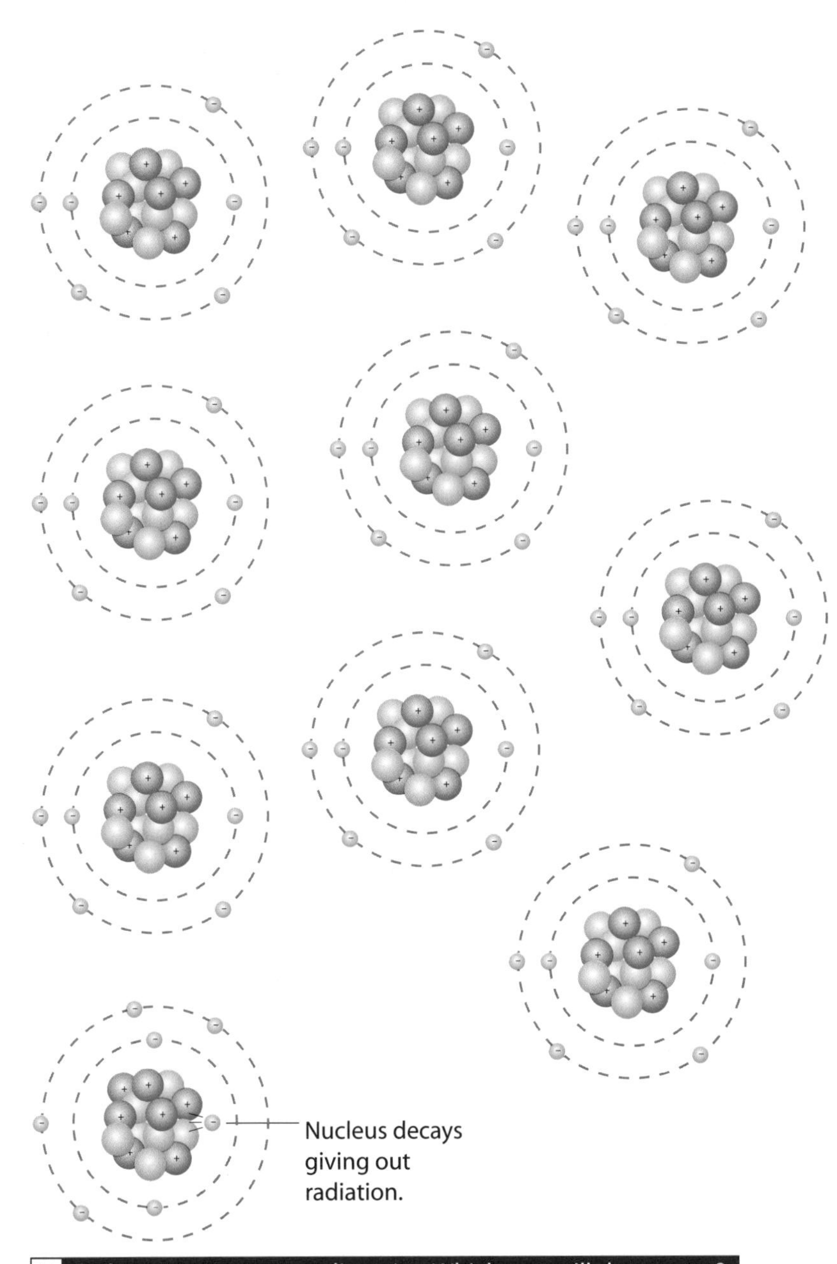

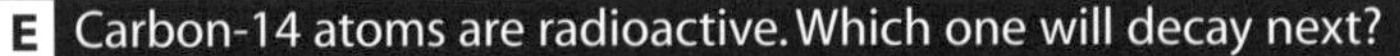

E Carbon-14 atoms are radioactive. Which one will decay next?

A substance with nuclei that decay is said to be **radioactive**. It is impossible to say when any one atom will decay but we can make good estimates about how many will decay each second from a large group of atoms. Some substances give out radiation from the nuclei of their atoms all the time, whatever is done to them.

7 If you had ten atoms of radioactive uranium, why couldn't you say when the next one will decay?

8 a Explain how isotopes are similar to one another.
  b Explain how isotopes are different from one another.

9 Sinclair said that to make his sample of radioactive uranium decay faster he would heat it. Why wouldn't Sinclair's plan work?

10 a Draw a radioactive atom of oxygen which has eight protons and nine neutrons in its nucleus.
  b Explain what you know about when the atom will decay.

Alpha, beta and gamma radiation

By the end of this topic you should be able to:

- describe the three types of nuclear radiation
- explain how each type of nuclear radiation comes about.

There are three main types of radiation given off by radioactive nuclei: **Alpha (α)** particles, **beta (β)** particles and **gamma (γ)** rays. Alpha and beta decay cause the atom to become a different element.

1 What do the following symbols mean?
a β **b** γ

2 What are the three types of radiation?

3 Which types of radiation cause an atom to become a different element?

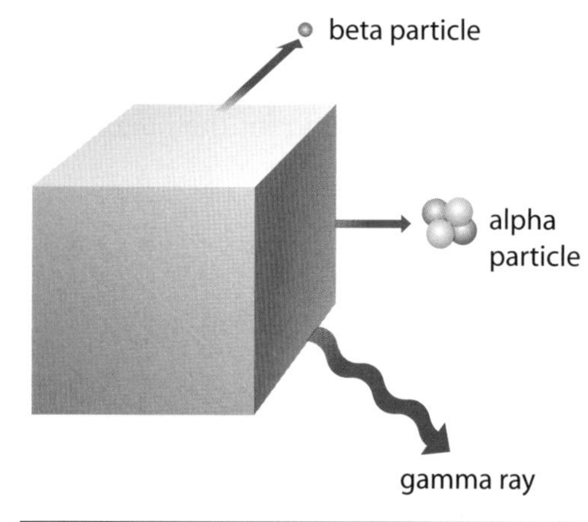

A Radioactive plutonium can give off all three types of radiation.

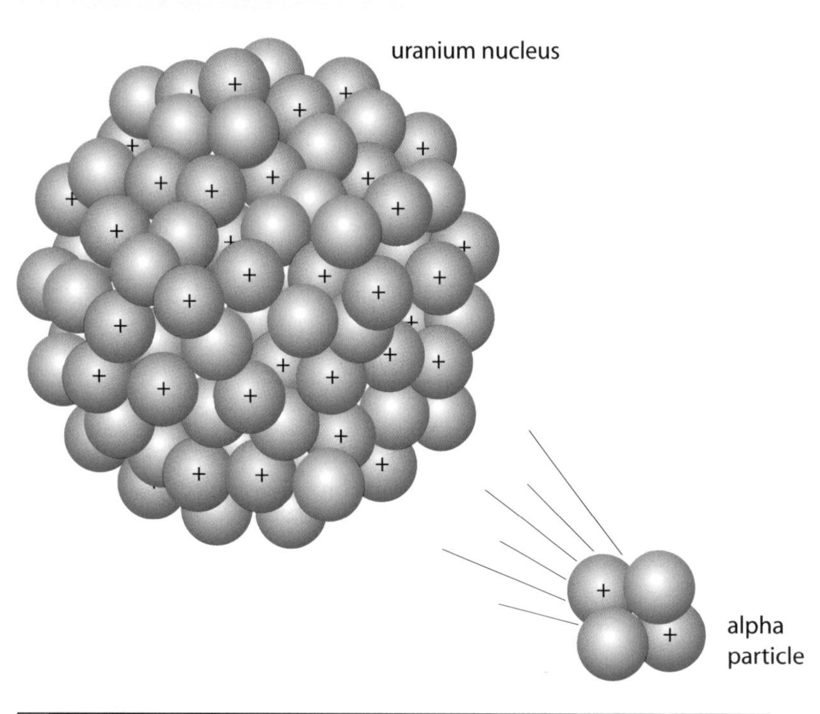

B A radioactive uranium-238 nucleus may give off an alpha particle.

An **alpha particle** is made from two protons and two neutrons. The four particles escape from the nucleus together and fly away from it. This makes alpha particles quite large. No electrons are involved. Alpha particles have a charge of +2.

When a nucleus gives off two protons, it then has a lower atomic number so it becomes a different element. For example, uranium begins with 92 protons. When it loses an alpha particle it only has 90 protons. This means it has become an atom of thorium.

4 How many of the following would you find in an alpha particle?
a electrons
b neutrons
c protons

5 Look at table C. An atom of protactinium decays by emitting an alpha particle.
a What is the atom's new atomic number?
b What element does the atom become?

Element	Atomic number
Actinium	89
Thorium	90
Protactinium	91
Uranium	92

C

A **beta particle** is a single fast-moving electron. Usually there are no electrons in a nucleus, but in beta decay, a neutron is changed into a proton and an electron. The proton stays in the nucleus but the electron flies off. The emitted electron is called a beta particle. It is very small and has a –1 charge.

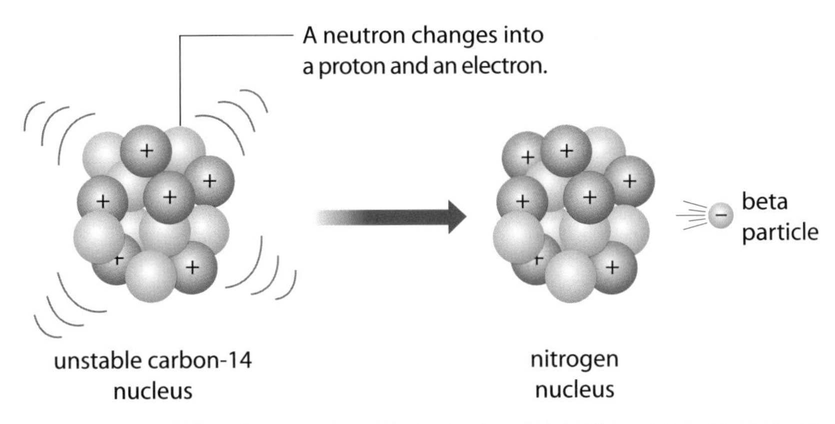

D Beta decay makes a neutron change into a proton and an electron. The electron as it flies off is a beta particle.

6 How many of the following would you find in a beta particle?
  a electrons
  b neutrons
  c protons

7 An atom of protactinium decays by emitting a beta particle.
  a What is the atom's new atomic number?
  b What element has the atom become? (*Hint*: look at table C.)

We looked at **gamma rays** as part of the electromagnetic spectrum. They are electromagnetic waves which come from unstable radioactive nuclei. As high-energy electromagnetic waves, gamma rays have no charge and are not made of particles. Gamma rays are only absorbed by very dense materials such as thick lead. Gamma decay does not change the structure of the atom.

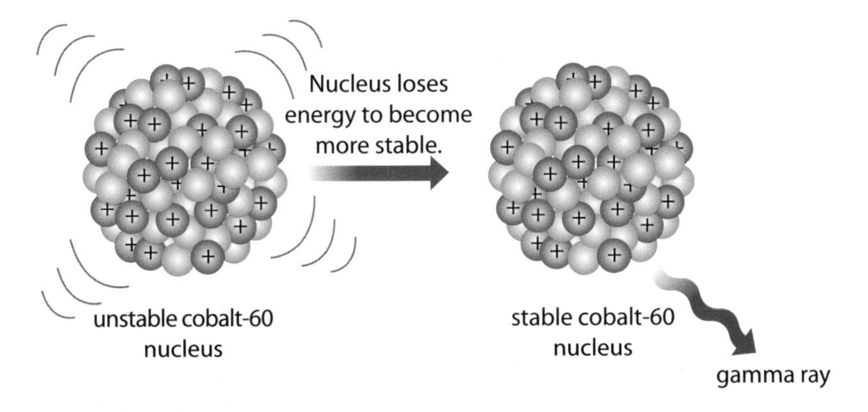

E Gamma decay takes energy away from a nucleus to make it more stable.

8 How many of the following would you find in a gamma ray?
  a electrons
  b neutrons
  c protons

9 An atom of protactinium decays by emitting a gamma ray.
  a What is the atom's new atomic number?
  b What element is the atom now?

10 Design a model to show the three types of radioactive decay. Draw your design and label it to explain how it shows:
  a alpha decay and the alpha particle
  b beta decay and the beta particle
  c gamma decay and the gamma ray.

P1b.13

Alpha radiation

By the end of this topic you should be able to:

- describe the uses of alpha radiation
- describe the hazards of alpha radiation
- evaluate when and how to reduce exposure to alpha radiation.

Blocking alpha particles

A What will block alpha particles?

Alpha particles are large and have a strong charge. They **ionise** other particles a lot. That is, they knock electrons off the particles and lose some energy. When an alpha particle loses enough energy it will stop. They are easily **absorbed** in only a few centimetres of air. They are also absorbed by very thin materials like paper or skin. This limits the uses of alpha particles.

Airports store a lot of fuels. They need very good fire alarm systems. One of the main uses of alpha sources is in smoke detectors. There is a small alpha source inside most of these. The alpha particles ionise air just outside the source. There is a small gap between the alpha source and an electric plate. The plate measures the electrons arriving from the ionisations. A constant stream of charged particles makes a constant current. Any smoke in the air gap absorbs some of the alpha particles and many of the ions. This makes the current drop. When the current drops enough, the siren sounds.

B How can a detector sense smoke?

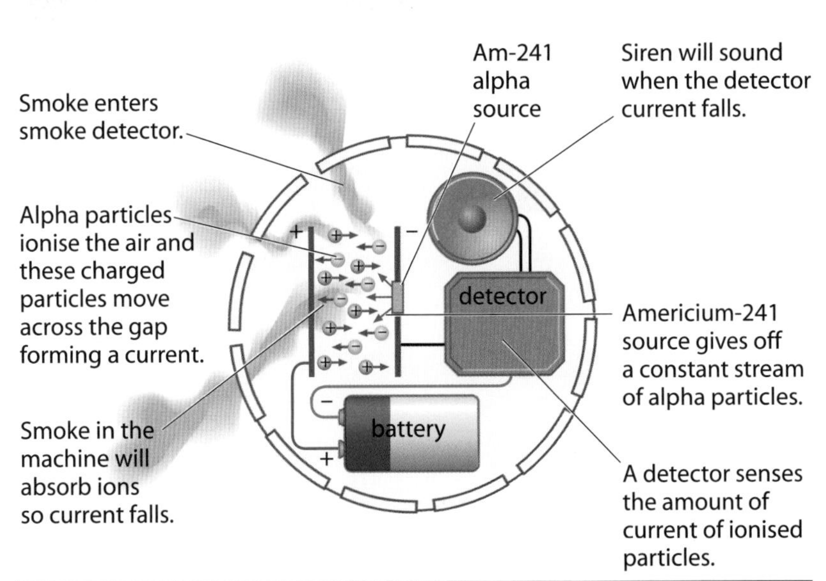

C A drop in alpha particle ionisation current shows there is smoke in the detector.

If alpha particles enter living cells, they do a lot of damage. This is because they ionise atoms in the cell. This may kill the cell, or could cause cancerous changes to occur. However, alpha particles are rarely dangerous since they are so easily absorbed by air before they reach you. Being just a few centimetres away from the source reduces the risk enough. Breathing in gases which are alpha sources, or getting them inside your body in some other way, can be very dangerous. This is because the cells absorb all the radiation.

Our biggest natural dose of radiation comes from radon gas. This is an alpha source and is common in all homes. It comes from the decay of the tiny amounts of uranium which are in all rocks and building materials, and the ground. The gas can get into houses. In most cases the amount is very small, but in a few cases the amounts are unsafe. These houses need special fans to pump the gas away.

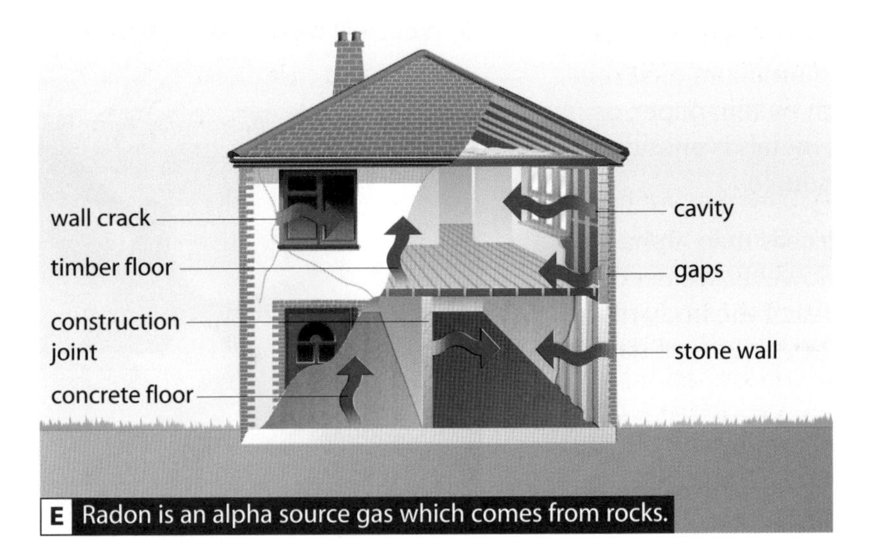

E Radon is an alpha source gas which comes from rocks.

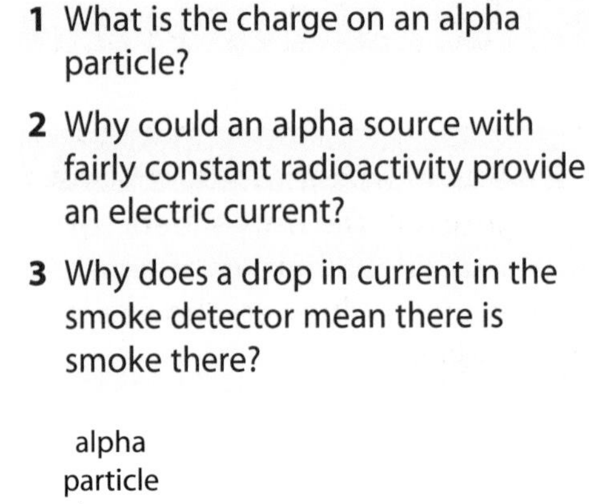

1 What is the charge on an alpha particle?

2 Why could an alpha source with fairly constant radioactivity provide an electric current?

3 Why does a drop in current in the smoke detector mean there is smoke there?

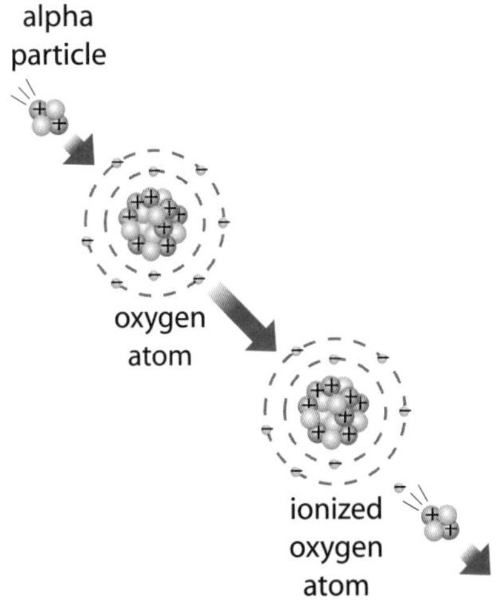

D Alpha particles are heavily ionising.

4 How can we protect ourselves from alpha sources in smoke detectors?

5 Why would a gas which gives off alpha particles be dangerous in the lungs?

6 Make a leaflet for homeowners to encourage them to install smoke detectors. Your leaflet should explain how smoke detectors work. Include diagrams. Explain how they are safe even though they use a radioactive source.

Beta radiation

By the end of this topic you should be able to:

- describe the uses of beta radiation
- explain the hazards of beta radiation
- evaluate when and how to reduce exposure to beta radiation.

Blocking beta particles

A A few millimetres of metal will block most beta particles.

Beta particles are single electrons. This means they are small and have a single negative charge. They ionise other particles less often than alpha particles. This means they can go further than alpha particles before they are stopped. A beta particle will be absorbed by a few millimetres of aluminium or several metres of air. They are not absorbed much by thin paper, or skin. A few millimetres thickness of most metals is enough to absorb virtually all beta particles from a source.

Beta particles are less damaging to living cells than alpha particles because they are less ionising. However, we need more protection from them as they can reach the body more easily. If beta particles enter living cells, they can cause damage and may lead to cancer.

1 What is the difference between the charges on alpha and beta particles?

2 What will block beta particles?

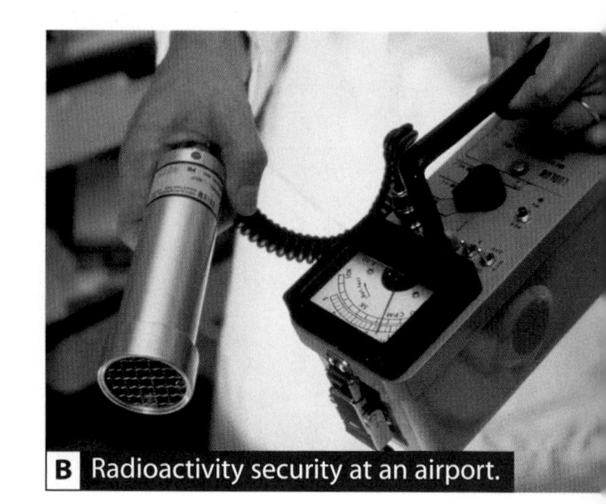

B Radioactivity security at an airport.

Carrying radioactive sources on a plane flight is dangerous and illegal. X-ray scanners cannot pick up radioactivity. To make sure no passengers have beta sources in their luggage, customs officers can check luggage without opening it by using a **Geiger-Müller counter**. The count rate for a beta source would be quite low as many beta particles would be absorbed before escaping from a case.

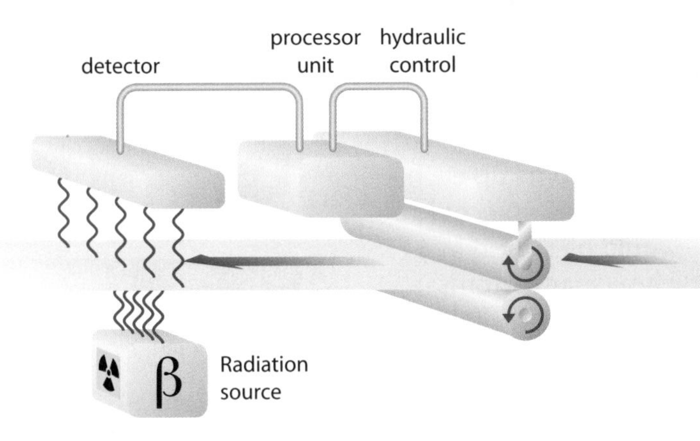

C Beta particles are used in controlling paper thickness in factories.

Some paper mills use a beta source to measure the thickness of their paper products. The beta source is placed below the paper and aimed upwards through the paper. A detector above measures the number of beta particles getting through the paper. If the paper gets thicker, it absorbs more beta particles and the count rate goes down. A computer sees the count drop and automatically increases the pressure between the rollers to make the paper thinner. This means the rolling machine is continuously controlled to keep the thickness even.

3 Why couldn't alpha particles be used in a paper mill?

4 Explain how the paper rollers will be changed if the paper becomes too thin.

A **tracer** is a chemical which can be followed to see where it moves. This idea was explained in Topic P1b.9. Sometimes a beta source can be used for a medical scan, but the absorption by the body means they are not as good.

Use of beta tracers

Beta tracers can be used for some natural studies. For example, silicon-32 is a beta source which is used as a tracer in geographical studies. The silicon-32 can be carried by water and by finding beta particles in a particular river, we know where the water has gone. This means we can follow the movement of water in an ecosystem.

5 If a beta tracer was used for a medical scan of a spine problem, why could you not necessarily rely on the results?

6 a How might using silicon-32 as a tracer in rivers be an environmental problem?
b How could this problem be reduced?

7 a Draw a diagram of a nucleus and show it decaying by beta decay.
b Label your diagram to explain the effects this decay would have on the nucleus, and the properties of the beta particle given off.
c Explain how you can stay safe when handling a beta source.

P P D P P

Radioactive half-lives

By the end of this topic you should be able to:

- explain what the half-life of a radioactive isotope means
- calculate the half-life of a radioactive isotope.

Radioactive decay is a random thing. It is not possible to predict when a particular nucleus will decay.

There is a probability that each nucleus will decay each second but, like throwing dice, you cannot say what will come up each time. If you threw the dice a thousand million times, you could say beforehand roughly how many sixes would be scored in total, but you could not be certain exactly how many. Atoms are so small that even a small amount of an element contains thousands of millions of them. So each second we could estimate, quite accurately, how many would decay. The only thing we couldn't predict would be which ones.

This idea of estimating random decay numbers leads to the idea of **half-life**. The half-life of a radioactive element is how long it takes for half the atoms in a group to decay. It can also be measured by finding the time it takes for the count rate of a sample to halve.

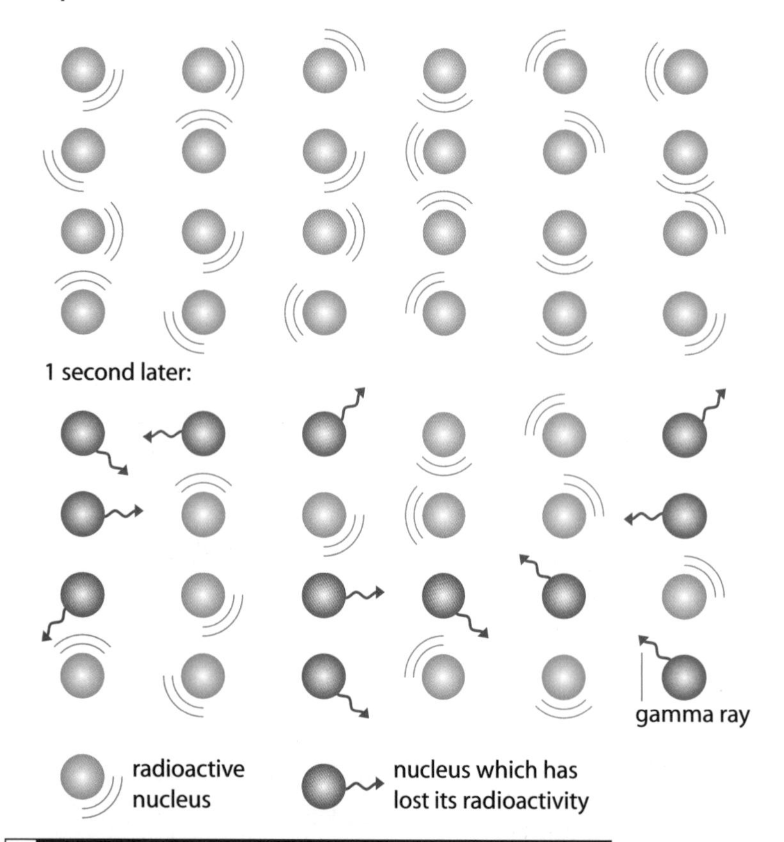

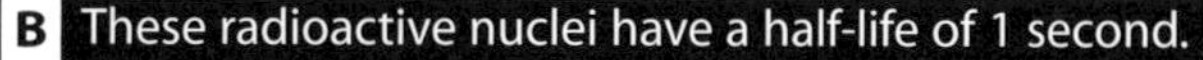

B These radioactive nuclei have a half-life of 1 second.

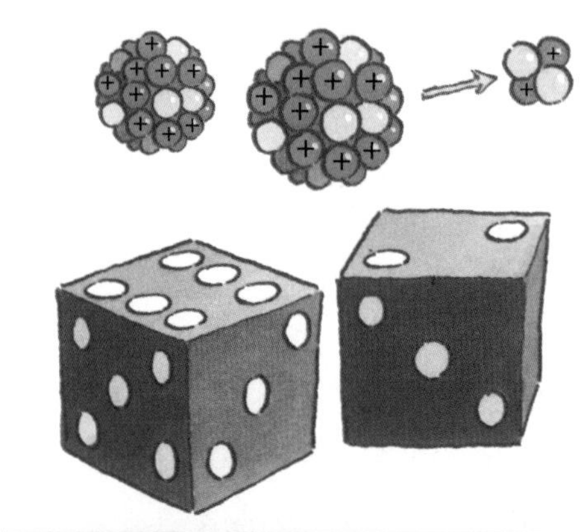

A When will you throw a six? When will a nucleus emit an alpha particle?

1. Why is radioactive decay like throwing dice?
2. If the chances of a radioactive nucleus decaying each second were one in six, how many would you expect to decay in 1 second from a group of 60 atoms?
3. In diagram B, the half-life is 1 second, so half of the atoms decay during the first second. How many nuclei would you expect to decay in the next second?
4. Strontium-90 decays with a half-life of 29 years. How many strontium-90 half lives have passed after:
   a 29 years **c** 87 years
   b 58 years **d** 145 years?
5. A sample of radioactive iron-59 contains one million atoms. The iron-59 has a half-life of 46 days. How many undecayed atoms will be left in the sample after:
   a 46 days **c** 138 days
   b 92 days **d** 184 days?

The half-life is different for each radioactive isotope. This is because it depends on how unstable the mix of protons and neutrons is. Half-lives range from a fraction of a second to billions of years.

Element	Half-life
Potassium-40	1.3 thousand million years
Carbon-14	5700 years
Strontium-90	29 years
Iron-59	46 days
Protactinium-234	71 seconds

C Some common half lives.

Measuring a half-life

D An experiment to measure half-life.

Working out a half-life

Looking at the numbers in a results table and estimating the half-life is not a very accurate way of finding the half-life, especially as radioactivity is random. One reading might be 'lucky' and be too high, or 'unlucky' and be too low. A much more accurate way is to draw a graph of the results, and draw a smooth curve between the points. This will average out the luck of the readings. The more readings you have, the more accurate the curve on your graph will be.

A graph of the count rate measured over time can be used to find half-life. Measuring from the graph, we can see how long it takes for the count rate to drop to half its starting value. This is the half-life.

6 If you account for background radiation, what was the actual first reading taken by the pupil in diagram D?

7 Estimate the half-life of protactinium from the results taken by the pupil in diagram D.

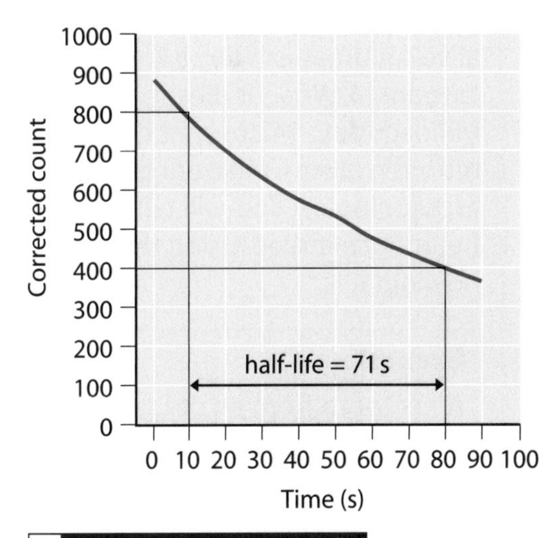

E Graph to find the half-life.

8 From graph E, what does 'corrected count' mean?

9 Using a graph to help, explain what it means to say that iron-59 has a half-life of 46 days.

P1b.16

Using radioactive half-lives

By the end of this topic you should be able to:

- explain how uses for radioactivity depend on knowing half-lives
- evaluate which radioactive isotopes would be best for certain uses because of their half-lives and the types of radiation they emit
- evaluate when and how to reduce exposure to nuclear radiations.

The half-life of a radioactive substance tells us how quickly its count rate will drop to half the start rate. If we know by what fraction the rate has dropped, we can tell how long it has been decaying.

Count rate (number per second)	Fraction of start rate	Number of half lives gone	Time since start (years)
10 000	start rate	0	start
5 000	1/2	1	5 700
2 500	1/4	2	11 400

A Knowing that carbon-14 has a half-life of 5700 years, archaeologists can work out the age of something.

Carbon-14 (C-14) is used to date archaeological objects. While alive, an organism has a known and constant amount of carbon-14. When it dies, it can't take in any more carbon, so the radioactive C-14 atoms it had will slowly decay to become nitrogen atoms. Measuring the amount of remaining C-14 in an archaeological find will tell us how many half-lives there have been since it died. Using this information, you can work out how old it is.

Example

A fox tooth from an old necklace has a count rate of 3000 counts per second (cps). A new fox tooth has a count rate of 12 000 cps. How old is the necklace?

count rate of new tooth: 12 000 cps
its count rate after 5700 years: 6000 cps
its count rate after 11 400 years: 3000 cps

So the necklace is about 11 400 years old.

1 Why would carbon-14 dating only be reliable for dating things that were once living?

2 Why could you not use C-14 dating to find the age of a living oak tree that is said to be 1000 years old?

3 Why would carbon-14 dating not work well for young objects, for example 10 years old?

Most countries ban the transport of very old relics from their country. To help fight against smuggling, customs officers sometimes use C-14 dating to confirm the age of a suspect object.

4 Why couldn't an ancient gold statue from Peru be dated using C-14 dating?

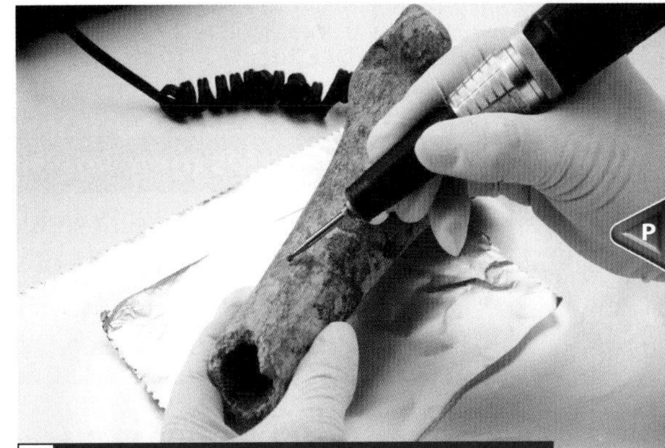

B Carbon-14 dating gives the age of archaeological treasures.

Medical scans using radioactive chemicals in the body can be dangerous. To reduce the danger, doctors use chemicals that have short half-lives. This means that they can carry out the scan using the radioactive chemicals, but the radioactivity quickly dies away to a safe level.

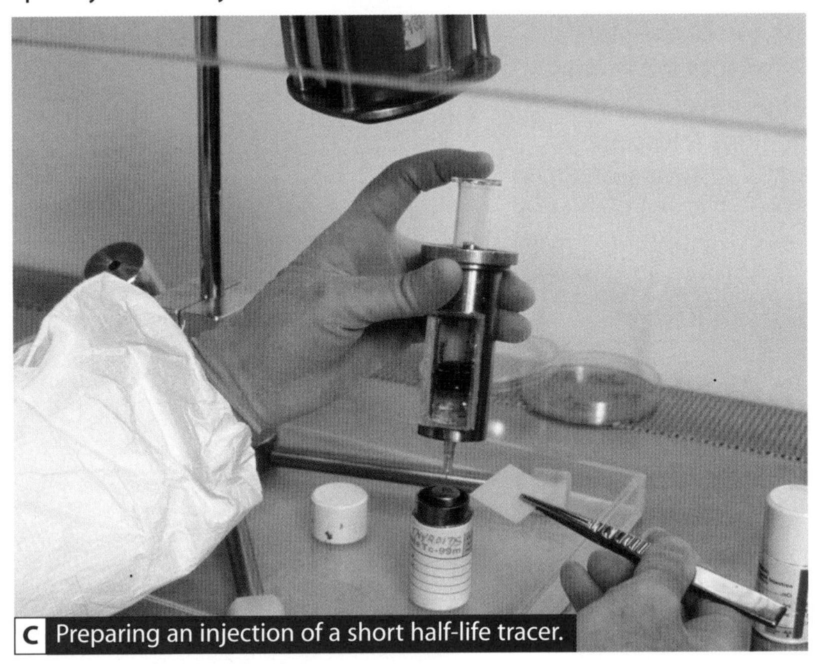

C Preparing an injection of a short half-life tracer.

Technetium-99m is a gamma source used in gamma camera scans. It has a half-life of only 6 hours. This means it has to be generated in the hospital laboratory and injected soon after. It also means that the radiation drops to a much lower, and safer, level after only 24 hours.

5 Why is using radioactive chemicals in the body for medical scans dangerous?

6 For technetium-99m:
- **a** how many half lives have passed in 24 hours
- **b** if the initial count rate was 18000 per second, what would it be 24 hours later
- **c** how long would it take to drop below 1000 per second?

The alpha source in an airport smoke detector needs to have a very long half-life. If it is too short, the activity will drop in a short time, and the detector's electronics will then think the alpha particles are being blocked by smoke and set off a false alarm.

7 Write a diary entry as if you are a customs officer who has discovered an old wooden pot in a passenger's luggage. You use carbon-14 to date the pot.
- **a** Explain what you check.
- **b** Explain why you check it in this way.
- **c** Explain what you would expect to find.

Telescopes galore

By the end of this topic you should be able to:

- describe which waves different types of telescope use
- compare the different types of telescope
- explain the advantages and disadvantages of telescopes based on Earth and in space.

You have seen that astronomers can use radio waves and gamma rays to study space. These astronomers use computers to turn the readings into visible-light pictures. We are more used to **telescopes** which we look through to see the stars with our eyes, using visible light. Seeing it for yourself is more exciting, but there are some limits to what our eyes can see. This limit is a disadvantage of using visible-light telescopes.

B Our atmosphere can block a lot of the visible light from stars.

Cloudy weather gets in the way of looking at stars because the light can't pass through clouds. Many parts of the electromagnetic spectrum can pass straight through the atmosphere.

1. Why is cloudy weather a problem for astronomers?
2. Suggest why the Hubble space telescope was placed in orbit above our atmosphere.
3. Name two different types of electromagnetic wave that can be used in astronomy.

Microwaves from space travel through the atmosphere, so some telescopes are designed to detect these and make pictures from the waves received. You can study space by looking for any part of the electromagnetic spectrum. Each type of electromagnetic wave has its own special properties and these make each of them useful for finding out different things about stars.

A The Hubble space telescope produces excellent pictures.

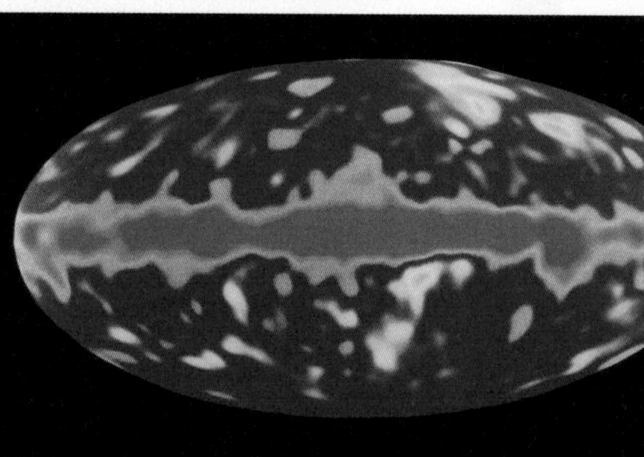

C A microwave view of the sky.

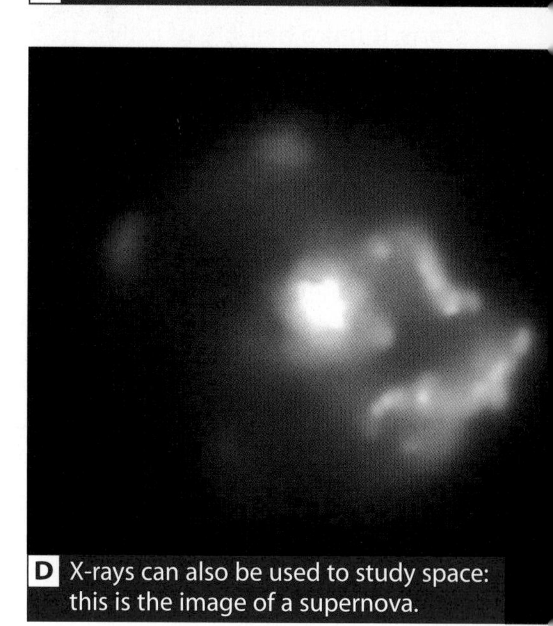

D X-rays can also be used to study space: this is the image of a supernova.

Infra red and UV are close enough to the wavelength of light that they can all be detected using similar telescopes. The **Hubble** telescope can see all three. Telescopes now have electronic sensors, like in a digital camera, that can show the picture on a screen. The pictures can be saved or changed using computers.

4 Suggest why radio telescopes are bigger than those for visible light? (*Hint*: think about the how the properties of electromagnetic waves change through the electromagnetic spectrum.)

5 Why do you think telescopes designed to see further need to be more sensitive?

Bad weather blocks light, but even on a clear night, the air in our atmosphere slightly distorts the light from **stars**. This means that what we see is not perfectly clear. Many observatories are built on high mountains where the air is thinner so their observations are better.

E The very large array interferometer in New Mexico, USA picks up metres long radio waves.

F Our atmosphere affects the light passing through it so stars aren't clear.

G Pic du Midi observatory in the Pyrenees mountains.

The next step was to build observatories in space to avoid the atmosphere altogether. The most famous of these is the Hubble space telescope. After some problems when launched, it is now taking spectacular photographs. Space-based telescopes have the advantages that they can operate day and night, and they can view the entire sky. The disadvantages of using telescopes based in space are that they are very expensive and harder to maintain than those on the ground.

6 Hubble cost about a billion pounds to build and launch. Do you think the money was worth spending? Give arguments for and against.

H The Eagle Nebula.

7 Write an item for a TV reporter who is telling viewers about the first pictures being sent back from the Hubble space telescope. Include a comparison with other types of telescope giving the advantages and disadvantages.

P D P P

The expanding Universe

By the end of this topic you should be able to:

- explain the effect on the frequency or wavelength observed when a wave source is moving
- explain what red-shift is, why we see it and how it changes
- evaluate the evidence the red-shift gives us about the Universe in the past and present.

A Sounds like the Doppler effect!

Have you ever noticed that a car, especially one with a siren like a police car or an ambulance, sounds different as it comes towards you compared to when it goes away from you? It happens because the sound waves are further apart if they are coming from something which is moving away from you. This is called the **Doppler effect**. It happens to all waves, including electromagnetic waves. We don't hear light though: if a galaxy is moving away from us, its colour changes. The colour shifts towards the red end of the visible spectrum, so it is called **red-shift**.

B Red-shifted light from galaxies tells us they are moving away.

The same effect happens across the whole electromagnetic spectrum. This means that what we might have expected to see as X-rays is red-shifted and becomes ultraviolet radiation when it reaches Earth. Red-shift tells us a galaxy is moving away. A bigger red-shift means the galaxy is moving away faster. We might find those same X-rays moved all the way into the visible region if they come from a faster moving galaxy.

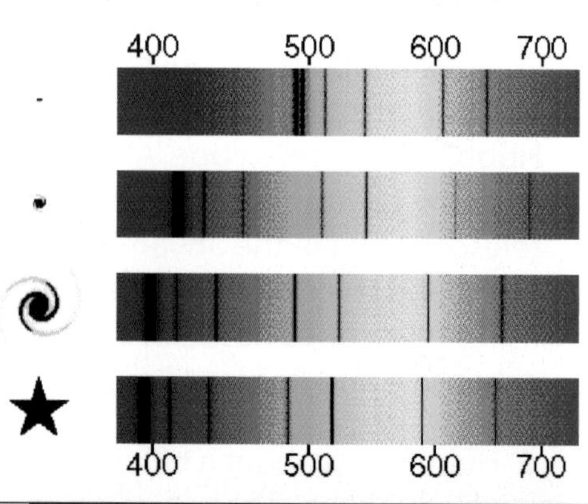

C Red-shift tells us about speed, and also distance to galaxies.

1 If you see light from a galaxy has a red-shift, what does this tell you?

2 Two galaxies show different red-shifts. How could you tell which one is moving away faster?

3 What might a blue-shift tell us?

The further a galaxy is from us, the faster it moves away. The further away a galaxy is, the bigger the red-shift we see. This means that, if we can measure its speed from the red-shift, then we can work out how far away a galaxy is. Measurements like this have helped us map out the **Universe**. The universe is everything we know exists.

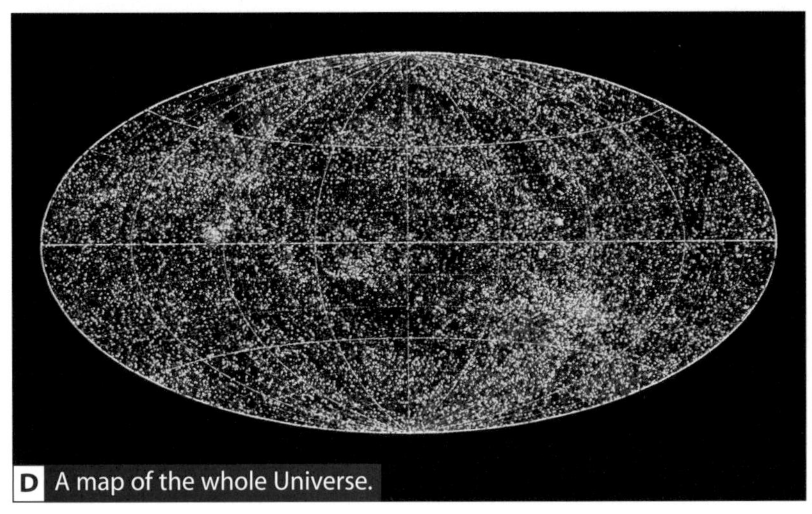

D A map of the whole Universe.

When we look at the galaxies in the sky, nearly all of them show a red-shift. This means almost everything in the Universe is moving away from us in all directions. This means that the Universe must be getting bigger.

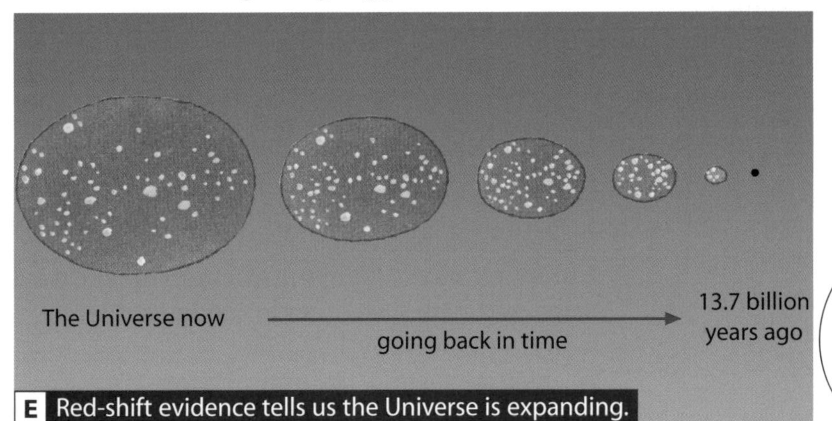

E Red-shift evidence tells us the Universe is expanding.

Red-shift evidence tells us that galaxies are moving apart from each other. If you were to make time go backwards, the Universe should get smaller. Eventually, everything in the Universe would crush back together into a single spot about 13.7 thousand million years ago. Playing time forwards again from there, a tiny dot suddenly explodes into the Universe we see today. This is the **Big Bang**. The red-shifts we see in the light from galaxies are strong evidence to support the Big-Bang theory.

6 What does the Big-Bang theory say?

7 What evidence is there for the Big-Bang theory?

4 Two galaxies show a different red-shift. How can you tell which one is further away?

5 What evidence do we have that the Universe is expanding?

F

8 The pupils in picture F can't quite get to grips with this topic. Write an explanation for them.

Assessment exercises

Part A

1 The Hubble telescope and the Chandra X-ray observatory are both in orbit around the Earth. Space-based telescopes have the advantage that:

- **a** They are cheaper than ground observatories.
- **b** They suffer no damage as space is a vacuum.
- **c** They can be easily maintained.
- **d** Their observations do not suffer atmospheric problems.

(1 mark)

2 The Hubble telescope uses visible light. Astronomers currently make observations using which parts of the electromagnetic spectrum?

- **a** Visible light only.
- **b** Visible light, X-rays and radio waves only.
- **c** Gamma rays, ultraviolet, infra red and visible light only.
- **d** All parts of the EM spectrum.

(1 mark)

3 All three telescopes in question 1 show that most distant galaxies have a red shift. What does the red shift tell us?

- **a** The galaxies are moving slowly.
- **b** The galaxies are orbiting the Milky Way.
- **c** The galaxies are moving away from us.
- **d** The galaxies are moving towards us.

(1 mark)

4 Misha's teacher showed her an experiment to see how much beta radiation was absorbed by different thicknesses of aluminium metal. The results of the experiment are given in the table below.

Thickness of aluminium (mm)	Beta activity (counts per second)
0.0	2600
0.5	2050
1.0	1550
1.5	1000
2.0	550
2.5	50

- **a** 0 counts per second
- **b** 2.5 mm
- **c** 2550 counts per second
- **d** 2600 counts per second

(1 mark)

5 For the experiment in Question 4, which of these statements is true?

- **a** 'Thickness of aluminium' is the independent variable and 'beta activity' is the dependent variable.
- **b** 'Thickness of aluminium' is the dependent variable and 'beta activity' is the independent variable.
- **c** Both are independent variables.
- **d** Both are dependent variables.

(1 mark)

6 Electromagnetic waves alter as their wavelength changes. Match the electromagnetic waves, listed **a–d**, to spaces **i–iv** in the table below.

- **a** gamma rays
- **b** infra-red
- **c** microwaves
- **d** ultraviolet

	increasing wavelength
i	
X-rays	
ii	
visible light	
iii	
iv	
radio waves	

(4 marks)

7 Many electromagnetic waves can be used in communication. Match the electromagnetic waves, listed **a–d**, with the communication uses **i–iv**.

a infra-red waves	**i** CCTV
b microwaves	**ii** optical fibres
c radio waves	**iii** ending information to satellites
d visible light	**iv** walkie-talkies

(4 marks)

8 Match the types of nuclear radiation, listed **a–d**, with their properties, **i–iv**.

a alpha	**i** The most penetrating.
b beta	**ii** Given off from a nucleus.
c gamma	**iii** Has the mass of an electron.
d all three types	**iv** The easiest to block.

(4 marks)

9 This is a diagram of a household smoke detector.

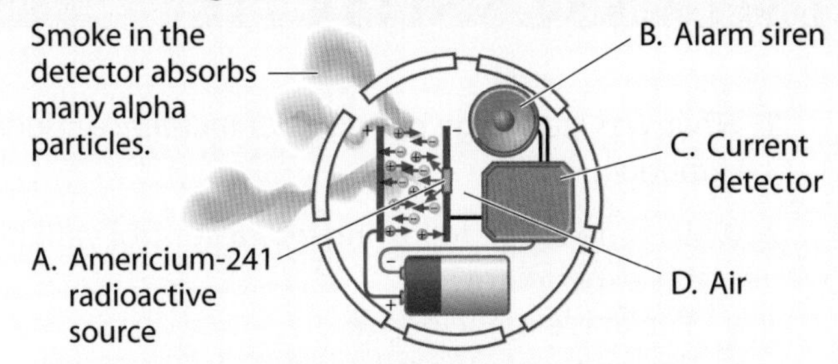

Match labels **a–d** to the spaces in sentences **i–iv**.

a ______ is ionized to form a current.
b ______ gives off alpha particles in a steady stream.
c ______ sounds when the detector current drops.
d ______ senses amount of air ionization. (*4 marks*)

10 Albert did an experiment with a light ray leaving a glass block.
He changed the angle of incidence from 10° up to 80°, going up in steps of 10° and for each angle of incidence, measured the angle of refraction at which the ray left the block.
Here are his results:

Angle of incidence (°)	Angle after (°)	Comments
10	15	Ray refracted out of glass
20	31	Ray refracted out of glass
30	49	Ray refracted out of glass
40	75	Ray refracted out of glass
50	50	Reflected inside glass
60	60	Reflected inside glass
70	70	Reflected inside glass
80	80	Reflected inside glass

His results show that he saw Total Internal Reflection when the angle of incidence was more than a special amount, called the Critical Angle.
Match the conclusions, **a–d**, with the results (**i–iv**) that allow you to make them.
Conclusions:

a The Critical Angle was at least 50°.
b You cannot tell the Critical angle exactly.
c There is no obvious pattern to the angle of a refracted ray.
d This experiment is not very **sensitive**.

Results:

i All the results.
ii Angles of incidence from 10° to 40°.
iii Angles of incidence of 40° and 50°.
iv Angles of incidence from 50° to 80°.

(*4 marks*)

Total (Part A) 25 marks

Part B

1 Write the answers that will complete this table showing the properties of nuclear radiations.

Name	Alpha	Beta	Gamma
Symbol	α	β	γ
Nature	**a**	electron	EM wave
Charge	2+	**b**	**c**
MassLarge	Tiny	**d**	

(*4 marks*)

2 Ank's teacher showed the class an experiment with radioactive Protactinium.
Firstly she recorded the background count rate: it was 0.5 counts per second. They she recorded the activity of the protactinium every ten seconds for four minutes.
Here are the results shown on a graph:

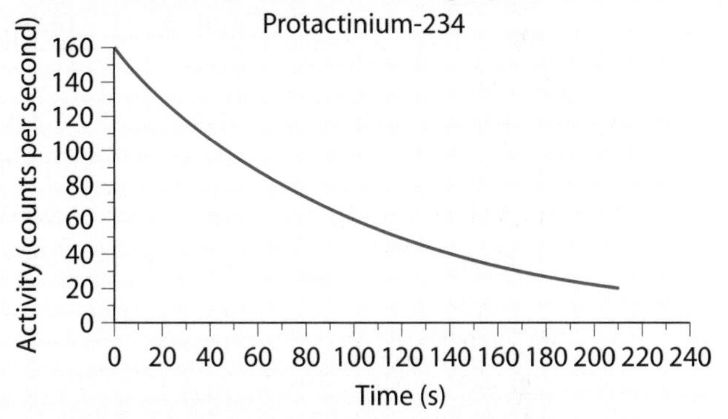

a When is the count rate 100? (*1 mark*)
b When is the count rate 50? (*1 mark*)
c What is the half life of protactinium? (*1 mark*)
d In drawing the graph, Ank ignored the background count reading. This gives a systematic error. What should he have done to account for the background count? (*1 mark*)

3 Nuclear radiations have different dangers.

a Why are alpha particle sources almost completely safe? (*1 mark*)
b How could an alpha source become dangerous? (*1 mark*)

4 Several studies on patients with brain tumours have suggested a link between heavy mobile phone use and these cancers. Mobile phone companies and the government's Health Protection Agency both agree that there is no risk and say we cannot trust these cancer studies.
Explain why a small number of people in the studies may not allow us to make valid conclusions. (*2 marks*)

Total (Part B) 12 marks

Investigative Skills Assessment

The *Lens-o-matic Company* makes lenses for telescopes. These need to be very accurate in how they bend and focus light rays. To check the accuracy of the telescopes, the company has samples of the materials for the lenses tested by two independent teams. The data below refers to a typical investigation. The two teams were each given

a a rectangular glass block, 10 cm by 6 cm by 2 cm

b a rectangular Perspex block with the same dimensions as the glass one.

They shone a ray of light into the block at various angles of incidence and recorded the angles of refraction.

The results are shown in tables A and B, below.

Angle of incidence (°)	Angle of refraction (°)	
	Team A	Team B
0.0	0.0	0.0
10.0	7.5	7.0
20.0	13.0	13.0
30.0	19.5	20.0
40.0	25.0	25.5
50.0	31.0	30.5
60.0	35.0	35.5
70.0	39.0	38.5
80.0	41.0	41.0

A Glass block.

Angle of incidence (°)	Angle of refraction (°)	
	Team A	Team B
0.0	0.0	0.0
10.0	6.5	6.5
20.0	13.0	13.0
30.0	19.0	19.5
40.0	25.0	25.0
50.0	35.5	31.0
60.0	35.0	35.5
70.0	39.0	38.5
80.0	41.0	41.0

B Perspex block.

Study the results tables carefully and then answer these questions.

1 What was the range of angles of incidence used? (*1 mark*)

2 Which one of the following was a control variable in this experiment?
- **a** The angle of incidence.
- **b** The angle of refraction.
- **c** The material of which the blocks were made.
- **d** The size of the blocks. (*1 mark*)

3 Which would be the best way of using the two sets of results for each material?
- **a** Add together the results for each material.
- **b** Find the mean for each angle for each material.
- **c** Find the median for each angle.
- **d** Find the mode of angles of refraction for each material. (*1 mark*)

4 Which would be the best way of presenting these results…
- **a** …to show how the angle of incidence changed with the angle of refraction for glass?
  - **i** A line graph of mean angles of refraction against angles of incidence.
  - **ii** A scatter graph of all glass results.
  - **iii** A bar chart with Team A's results in red and Team B's in blue.
  - **iv** A pie chart. (*1 mark*)
- **b** …to show how the results were different for glass and Perspex?
  - **i** A line graph of mean angles of refraction against angles of incidence.
  - **ii** A scatter graph of all results.
  - **iii** A bar chart with Team A's results in red and Team B's in blue.
  - **iv** A pie chart. (*1 mark*)

5 There is one anomalous result in the tables. In which table does it occur and what is its value? (*1 mark*)

6 Why do you think that for an angle of incidence of 0.0° both teams got exactly the same results? (*1 mark*)

7 Why did the company employ two independent teams to do these experiments?
- **a** To improve accuracy.
- **b** To improve sensitivity.
- **c** To improve precision.
- **d** To improve reliability. (*1 mark*)

8 a Complete this sentence comparing the results for glass and Perspex. *The angles of refraction for the glass and Perspex block were…* (*1 mark*)

b What do the results for glass and Perspex tell you about choosing which material to use for telescope lenses? (*1 mark*)

9 What do the results suggest about building a lens with an angle of refraction of 50°? (*1 mark*)

10 How could the *Lens-o-matic Company* improve this experiment in order to make their advertising more convincing to astronomers who buy telescope lenses?

a Include other materials for comparison.

b Only use one team to simplify the results.

c Use a more sensitive instrument to measure the angles to 0.1°.

d Reverse the ray direction, so the light is leaving the blocks. (*1 mark*)

Total = 11 marks

Glossary

absorb Take in, or soak up.
alpha (α) particle Radioactive particle which is two protons and two neutrons combined.
analogue signals Information that can vary at many levels.
atomic number The number of protons in the nucleus of an atom of an element.
atom The smallest part of an element.
background radiation Radiation occurring naturally in the environment.
beta (β) particle Radioactive particle which is a fast-moving electron.
Big Bang A theory about the origin of the Universe which says that everything exploded outwards from a single point.
decay Change in a nucleus which gives off radiation.
digital signals Information which can have only two states, on or off.
Doppler effect Change in the wavelength of a wave caused by movement of the wave source.
electromagnetic radiation A form of energy transfer, including radio waves, microwaves, infra red, visible light, ultraviolet, X-rays and gamma rays.
electromagnetic spectrum The groupings of electromagnetic radiation in order according to frequency/wavelength.
electron Negatively charged particle which circles an atom's nucleus.
frequency The number of wave vibrations per second.
gamma (γ) ray High-frequency electromagnetic radiation given off by a radioactive nucleus.
gamma-ray astronomy Observing the sky in the gamma ray part of the electromagnetic spectrum.
Geiger-Müller counter Machine to detect radioactivity.
half-life The time taken for the number of radioactive particles to fall by half.
hertz (Hz) The common unit for measuring frequency, waves per second.
Hubble telescope Space-based telescope observing in UV, IR, and visible light.
infra red radiation Electromagnetic radiation that we can feel as heat. It has a longer wavelength than visible light, but a shorter wavelength than microwaves.
interference Random, unwanted variations in a signal.
ionise Add or remove electrons from an atom to leave it charged.
isotopes Atoms with the same atomic number but different mass numbers.
lead Very dense metal capable of absorbing much X-ray energy.
mass number The number of protons plus the number of neutrons in an atom of an element.
microwaves Electromagnetic radiation that has a longer wavelength than infra red but shorter than radio waves.
neutron A neutral particle found in the nucleus of an atom of an element.
noise Random, unwanted variations in a signal.
nucleus (plural **nuclei**) Central part of an atom, made up of protons and neutrons.
optical fibre Glass or plastic strand which can direct light or infra red radiation.
proton A positively charged particle found in the nucleus of an atom of an element.
radar Acronym: **RA**dio **D**etection **A**nd **R**anging. System for detecting the positions of objects by reflecting radio waves off them.
radio waves Electromagnetic radiation with the longest wavelength and lowest frequency.
radioactive Something that gives out dangerous radiation when it decays.
radioactive nucleus Unstable centre of an atom, which gives off radiation to become more stable.
radiotherapy Cancer treatment in which a patient is dosed with gamma radiation to kill the cancer cells.
red-shift Doppler effect in electromagnetic radiation from distant galaxies.
reflect When something bounces off an object, such as light bouncing off a mirror.
stable A nucleus which will not decay.
star A body in space which gives off light as a result of nuclear reactions inside it.
telecommunication Communication over a distance by a cable or wire.
telescopes Machine for observing bodies in space.
total internal reflection At more than critical angle, light trying to leave a more dense medium, like glass, is reflected back inside with no refraction.
tracer Chemical which can be detected to observe its movements.
transmit Allow to pass through.
ultraviolet (UV) Electromagnetic radiation that has a shorter wavelength than visible light but longer than X-rays.
Universe The region that contains everything that exists.
vacuum A place where there is no matter at all, e.g. space.
visible light Electromagnetic radiation that can be sensed by human eyes.
wavelength (λ) the distance between the same point on one wave to the next.
X-rays Electromagnetic radiation that has a shorter wavelength than UV but longer than gamma rays.

Index

Pearson Education
Edinburgh Gate
Harlow
Essex
CM20 2JE
UK
www.longman.co.uk

First published 2006
Second impression 2006

ISBN-13: 978-1-4058-3327-1/ ISBN-10: 1-4058-3327-0
Printed in Great Britain by CPI Bath

Design and production	Roarr Design
Illustration	Oxford Designers & Illustrators Ltd
Picture research	Kay Altwegg
Indexer	Indexing Specialists (UK) Ltd

The publisher's policy is to use paper manufactured from sustainable forests.

Acknowledgments

The publishers are grateful to the following for their collaboration in reviewing this book.

Jo Clark B.Sc., Head of Science, The Ridings School, Halifax, West Yorkshire; Deborah Dury, Science Teacher, Wrockwardine Wood Arts College, Telford, Shropshire; Stuart Fink, Head of Science, King Solomon High School, Redbridge, Ilford; Joyce Gustard, Head of Science, St Gabriel's School, Newbury, Berkshire; Martin Jago, Manager of Science, John Spendluffe Technology College, Alford, Lincolnshire; Andy Makucewicz, Head of Science, Trinity School, Belvedere, Kent; Mr J A Shropshire, Head of Science, All Hallows High School, Macclesfield Cheshire; Claire Thornton, Science Teacher, Sandhill View School Sunderland; Dr Mike Viccary, Head of Science, Bradford Christian School.

Photo Acknowledgements

The publishers are grateful to the following for their permission to reproduce copyright photographs.

Action Plus: pg15 (Leo Mason), pg18, pg20, pg26(l), pg34 (Glyn Kirk), pg19(mt), pg22, pg27(all) (Neil Tingle); **Alamy:** pg24(t) (Adrian Muttitt), pg24(m) (Images of Africa Photobank), pg26(r) (blickwinkel), pg30 (Image Source), pg33(b) (Brand X Pictures), pg46(r) (image100), 47(t) (Danita Delimont), pg49(t), pg51(tr) (david tipling), pg50(b) (Andrew Harrington), pg51(mr) (SuperStock), pg51(mr) (Visual&Written SL), pg52, pg73(tr), pg135, pg148(l), pg195(l) (Photofusion Picture Library), pg54(l), pg137(l), pg139 (Holt Studios International Ltd), pg68(t), pg80(bl) (David R. Frazier Photolibrary, Inc), pg68(b), 109(tr) (david sanger photography), pg71(b) (David Noton Photography), pg72, pg216 (Chris Howes/Wild Places Photography), Pg73(br) (fotolincs), pg77 (David Woodfall), pg78 (ImageState), pg80(br) (Mark Sykes), pg88(t) (Paul Glendell), pg89(tl) (Agripicture Images), pg89(bl), pg101, pg189(t) (Leslie Garland Picture Library), pg90(t) (Nic Cleave Photography), pg91(t) (Guy Edwardes Photography), pg100(bl) (Karin Duthie), pg100(br) (South West Images Scotland), pg103(t) (Bill Heinsohn), pg105(tr) (Banana Stock), pg105(bl), 109(br), 130(b) (foodfolio), pg106(bl) (Bill Marsh Royalty Free Photography), pg106(br) (Design Pics Inc), pg107 (l) (Mark Baigent), pg109(bl) (Paul Broadbent), pg112(t), pg192 (Peter Bowater), pg112(b) (AGStockUSA, Inc.), pg113(l) (Bildagentur Hamburg), pg114(br) (TOM MARESCHAL), pg114(l) (Motoring Picture Library), pg117(l) (Davo Blair), pg117(b) (Dinodia Images), pg119(t) (Danita Delimont), pg119(b) (Dan Atkin), pg132(b) (ACE STOCK LIMITED), pg133(tr) (Jef Maion/Nomads'Land-www.maion.com), pg133(br) (image100), pg134(l) (Robert Brook), pg147(b) (Liam Bailey), pg157(t) (Image Source), pg157(b) (Digital Archive Japan), pg178 (isifa Image Service s.r.o.), pg180 (Robert Harding Picture Library Ltd), pg182(br) (STOCK IMAGE/PIXLAND), pg183(br) (bilderlounge), pg183(t) (Mark Harwood), pg183(bl) (Stephen Oliver), pg189(b) (Brian Hamilton), pg194 (Mark Sykes), 196(b) (Powered by Light/Alan Spencer), pg198, pg220(br) (David Hoffman Photo Library), pg213 (plainpicture GmbH & Co. KG), pg215(t) (Gari Wyn Williams), pg217(t) (Doug Houghton), pg221(tl) (Steve Allen), pg223(l) (Bubbles Photolibrary), pg225(r) (Mike Hill), pg225(l) (Janine Wiedel Photolibrary), pg231 (Blackout Concepts); **Anthony Blake:** pg140(t) (Maximillian Stock Ltd); **Ardea London Ltd:** pg50(t) (Chris Harvey), pg65 (John Mason), pg66(l) (Jean Michel Labat); **BAA Aviation Photo Library:** pg209(tl), pg220(l), pg221(bl); **Bjørn Rørslett-NN/Nærfoto:** pg221(l); **Construction Photolibrary:** pg87(all), pg90(b) (Grant Smith), pg89(tr), pg91(b), pg106(tr) (Chris Henderson), pg105(tl) (David Stewart-Smith), pg106(tl) (Mike Cooper), pg113(r) (David Stewart-Smith), pg130(m) (DIY Photolibrary), pg184(t) (Jean-Francois Cardella); **Corbis:** pg36(t) (Royalty-Free), pg36(b) (Lester V. Bergman), pg103(b) (Paul A Souders), pg200 (Beirne Brendan/Corbis Sygma), pg219(r) (David Michael Zimmerman), pg220(tr) (Reuters); **EMPICS Sports Photo Agency:** pg182(l); **FLIR Systems:** pg219(l); **FLPA:** pg48(b) (MICHAEL & PATRICIA FOGDEN/Minden Pictures), pg73(l) (John Hawkins); **Food Features:** pg24(b), pg25, pg110(t), pg125, pg130(mt)(mb), pg133(l), pg134(r), pg136(r), pg138(tr), pg138(bl), pg141(all), pg142, pg146, pg152, pg179, pg191, pg217(b); **GeoScience Features Picture Library:** pg88(ml)(mr)(b); **Getty Images:** pg95(br) (Time & Life Pictures), pg100(t) (Hulton Archive); **Hockerton Housing Project:** pg167(l), pg170(t), pg173, pg202; **Illustration by Karren Carr:** pg62(br); Kimbleton Fireworks: pg186; **Mari Tudor-Jones:** pg9, pg10, pg19 (all), pg45, pg61(t), pg80(tl), pg138(tl), pg144, pg155, pg167(r), pg174(l), pg196(t), pg209(m), pg214(t), pg214(m), pg221(tr), pg230; **NASA:** pg62(tr), pg242(tr)(1997), (CXC/SAO), pg243(br) (ESA, Hubble Heritage (STScI/AURA)); **naturepl.com:** pg48(t) (Jason Venus), pg58(t) (Tim Shephera), pg75(l) (Niall Benvie), pg75(r) (Geoff Simpson), pg80(tr) (Martha Holmes); **New Media Ltd:** pg95(tr); **NHMPL:** pg62(bl) (De Agostini); **NHPA:** pg49(b) (T KITCHIN & V HURST), pg51(l) (ROD PLANCK), pg56(t), pg57 (GEORGE BERNARD), pg63 (DANIEL HEUCLIN), pg66(r) (YVES LANCEAU), pg74 (JOHN SHAW) pg199(b) (ALAN WILLIAMS); **Image courtesy of NRAO/AUI:** pg243(tr); **The Ohio State University Radio Observatory/ NAAPO/www.bigear.org:** pg215(b); **Panos:** pg71(t) (Stefan Boness); **Redferns:** pg184(b), pg188(t) (MICK HUTSON); **Rex Features:** pg108(l), pg109(mr); **Roslin Institute:** pg59; **RWE npower:** pg203(l); **Samantha Hall:** pg47(r); **Science Photo Library:** pg32(r), pg53, pg61(b) (Mauro Fermariello), pg32(l) (Stanley B. Burns, MD & The Burns Archive N.Y.), pg33(tr) (Hybrid Medical Animation), pg33(br) (Photo Insolite Realite), pg33(mr) (Dr.Linda Stannard, UCT), pg33(tl) (Ton Kinsbergen), pg33(m) (Peter Menzel), pg33(ml) (Kent Wood), pg46(l) (Geoff Kidd), pg56(b) (DAVID SCHARF), pg58(b) (SINCLAIR STAMMERS), pg67, pg69 (Philippe Plailly/Eurelios), pg76 (M-SAT. Ltd), pg92, pg108(r) (SHEILA TERRY), pg95(l) (ANDREW LAMBERT PHOTOGRAPHY), pg102 (MARTYN F. CHILLMAID), pg104 (MAXIMILIAN STOCK LTD), pg107(r) (PASCAL GOETGHELUCK), pg114(tr), pg137(r) (ANDREW LAMBERT PHOTOGRAPHY), pg117(t) (Adam Hart-Davis), pg118, pg160(t), 197(b), pg234 (CORDELIA MOLLOY), pg133(mr) (AJ PHOTO), pg140(b) (Charles D Winters), pg147(t) (DAMIEN LOVEGROVE), pg148(r) (Andrew McClenaghan), pg160(b) (MICHAEL MARTEN), pg174(r) (TED KINSMAN), pg176(l) (ADRIENNE HART-DAVIS), pg176(r) (PAUL WHITEHILL), pg195(r) (NOVOSTI), pg197(tl), pg199(t) (MARTIN BOND), pg209(r) (US DEPARTMENT OF ENERGY), pg209(bl) (EUROPEAN SPACE AGENCY), pg212 (DAVID PARKER), pg214(b) (ROBIN SCAGELL), pg223(r) (PETER SKINNER), pg227(r) (BSIP/CAVALLINI JAMES, EUROPEAN SPACE AGENCY / M. REVNIVTSEV (IKI/ MPA)), pg236 (HANK MORGAN), pg241(t) (JAMES KING-HOLMES), pg241(b) (DAVID PARKER), pg242(l) (MAGRATH PHOTOGRAPHY), pg242(mr) (NASA GSFC), pg243(mr) (J-C CUILLANDRE / CANADA-FRANCE-HAWAII TELESCOPE), pg243(l) (PHILIPPE PSAILA), pg244(l) (CHRIS BUTLER), pg245 (MAX-PLANCK-INSTITUT FUR EXTRATERRESTRISCHE PHYSIK); **info@scottishviewpoint.com:** pg203(r); **Still Pictures:** pg70 (Shehzad Noorani), pg116 (Peter Frischmuth), pg136(l); **Sue Kearsey:** pg47(l); **Warren Photographic:** pg54(r), pg55; **www.astro.ucla.edu:** pg244(r); **www.Colourchange.com:** pg132(t); **www.element-collection.com/Theodore W. Gray:** pg157(m); **www.glofish.com:** pg60; **www.thinkroadsafety.gov.uk:** pg13; **www.tidalstream.co.uk:** pg197(tr)

The following photographs were taken on commission © **Pearson Education Ltd** by:
Trevor Clifford: pg93, pg96(all), pg98(all), pg110(b), pg126, pg127, pg128, pg130(t), pg131 (Win Health Ltd), pg168(all), pg170(b), pg182(tr), pg187, pg188(b), pg190(Defra), pg210, pg222(tl)(bl)

Front cover photos:
Main image: Mari Tudor-Jones
Inset: (top) Mari Tudor-Jones; (middle) Sascha/Getty Images (Photonica); (bottom) Luis Castaneda Inc/Getty Images (ImageBank).

Every effort has been made to trace the copyright holders and we apologise in advance for any unintentional omissions. We would be pleased to insert the appropriate acknowledgement in any subsequent edition of this publication.

Licence Agreement: *AQA GCSE Science CD-ROM*

Warning:
This is a legally binding agreement between You (the school) and Pearson Education Limited of Edinburgh Gate, Harlow, Essex, CM20 2JE, United Kingdom ('PEL').

By retaining this Licence, any software media or accompanying written materials or carrying out any of the permitted activities You are agreeing to be bound by the terms and conditions of this Licence. If You do not agree to the terms and conditions of this Licence, do not continue to use the CD and promptly return the entire publication (this Licence and all software, written materials, packaging and any other component received with it) with Your sales receipt to Your supplier for a full refund.

AQA GCSE Science CD-ROM consists of copyright software and data. The copyright is owned by PEL. You only own the disk on which the software is supplied. If You do not continue to do only what You are allowed to do as contained in this Licence you will be in breach of the Licence and PEL shall have the right to terminate this Licence by written notice and take action to recover from you any damages suffered by PEL as a result of your breach.

Yes, You can:

1. use or install *AQA GCSE Science CD-ROM* on Your own personal computer as a single individual user:

No, You cannot:

1. copy *AQA GCSE Science CD-ROM* (other than making one copy for back-up purposes);
2. alter *AQA GCSE Science CD-ROM*, or in any way reverse engineer, decompile or create a derivative product from the contents of the database or any software included in it:
3. include any software data from *AQA GCSE Science CD-ROM* in any other product or software materials;
4. rent, hire, lend or sell *AQA GCSE Science CD-ROM*;
5. copy any part of the documentation except where specifically indicated otherwise;
6. use the software in any way not specified above without the prior written consent of PEL

Grant of Licence:
PEL grants You, provided You only do what is allowed under the Yes, You can table above, and do nothing under the No, You cannot table above, a non-exclusive, non-transferable Licence to use *AQA GCSE Science CD-ROM*.

The above terms and conditions of this Licence become operative when using *AQA GCSE Science CD-ROM*.

Limited Warranty:
PEL warrants that the disk or CD-ROM on which the software is supplied is free from defects in material and workmanship in normal use for ninety (90) days from the date You receive it. This warranty is limited to You and is not transferable.

This limited warranty is void if any damage has resulted from accident, abuse, misapplication, service or modification by someone other than PEL. In no event shall PEL be liable for any damages whatsoever arising out of installation of the software, even if advised of the possibility of such damages. PEL will not be liable for any loss or damage of any nature suffered by any party as a result of reliance upon or reproduction of any errors in the content of the publication.

PEL does not warrant that the functions of the software meet Your requirements or that the media is compatible with any computer system on which it is used or that the operation of the software will be unlimited or error free. You assume responsibility for selecting the software to achieve Your intended results and for the installation of, the use of and the results obtained from the software.

PEL shall not be liable for any loss or damage of any kind (except for personal injury or death) arising from the use of *AQA GCSE Science CD-ROM* or from errors, deficiencies or faults therein, whether such loss or damage is caused by negligence or otherwise.

The entire liability of PEL and your only remedy shall be replacement free of charge of the components that do not meet this warranty.

No information or advice (oral, written or otherwise) given by PEL or PEL's agents shall create a warranty or in any way increase the scope of this warranty.

To the extent the law permits, PEL disclaims all other warranties, either express or implied, including by way of example and not limitation, warranties of merchantability and fitness for a particular purpose in respect of *AQA GCSE Science CD-ROM*.

Governing Law:
This Licence will be governed and construed in accordance with English law.